CAMBRIDGE MONOGRAPHS IN
EXPERIMENTAL BIOLOGY
No. 5

EDITORS:

M. ABERCROMBIE, P. B. MEDAWAR
GEORGE SALT (*General Editor*)
M. M. SWANN, V. B. WIGGLESWORTH

THE WATER RELATIONS OF TERRESTRIAL ARTHROPODS

THE SERIES

Other volumes in preparation

THE WATER RELATIONS OF TERRESTRIAL ARTHROPODS

BY

E. B. EDNEY, D.Sc.

Professor of Zoology in the
University College of Rhodesia and Nyasaland,
formerly Reader in Entomology in the
University of Birmingham

CAMBRIDGE
AT THE UNIVERSITY PRESS
1957

PUBLISHED BY
THE SYNDICS OF THE CAMBRIDGE UNIVERSITY PRESS

London Office: Bentley House, N.W.1
American Branch: New York

Printed in Great Britain by
Western Printing Services Ltd, Bristol

CONTENTS

ACKNOWLEDGEMENTS

Permission to refer to unpublished work by the following is gratefully acknowledged: Dr J. L. Cloudsley-Thompson, Dr M. W. Holdgate, Mr A. R. Mead-Briggs. I am also indebted to Dr J. W. L. Beament, Dr M. W. Holdgate and Professor V. B. Wigglesworth, whose willingness to discuss and advise on matters of common interest has been of great value.

E. B. E.

UNIVERSITY COLLEGE OF
RHODESIA AND NYASALAND

CHAPTER I

INTRODUCTION

A MONOGRAPH concerned with water relations must define its scope rather arbitrarily, for the term has been variously interpreted according to the interests of the user.

Historically, one of the main stimuli for the study of water relations came from ecology. Many facts about the relation between climate and the abundance, distribution and activities of arthropods—insects in particular—were collected during the early years of this century, but they were largely unorganized, and by 1931, Uvarov's classic review on 'Insects and climate' referred frequently to the need for more fundamental physiological knowledge. Since then progress has been made along these lines, and it is one aspect of this work, namely the physiology of water balance (interpreted fairly widely), which forms the subject-matter of the present book.

Transpiration, its control and its effects will be treated in some detail, together with a consideration of the structure and properties of cuticles, particularly as regards water-proofing. Respiration and excretion will also be considered so far as they contribute to water loss. Water uptake will be discussed, and this will include consideration of the role of metabolic water, and of absorption of water and water vapour from the environment. Internal regulation of water and electrolytes will clearly form another main field of inquiry. Finally the significance of water in relation to body temperature and thermal resistance will be considered.

THE SATURATION DEFICIT 'LAW'. In much of the earlier work on water relations the effects of physical factors, particularly temperature and humidity, upon development and longevity formed the centre of interest, and attempts were made to find a simple relation between these variables in the hope of reducing the mass of ecological observations to order and also of predicting fluctuations in insect numbers. The drying power of the air was clearly a factor of great significance, and if a combination of the effects of temperature and humidity could be expressed in these terms a considerable simplification would be achieved.

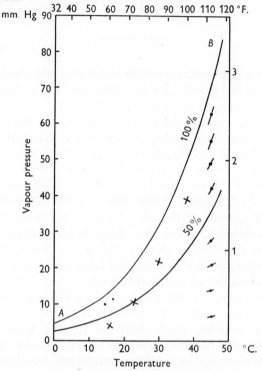

FIG. 1. The curve *A–B* shows the saturation vapour pressure of water in air at various temperatures. The curve for 50% relative humidity is also drawn, and other relative humidities are indicated. The crosses all represent saturation deficits of 10 mm. Hg at different temperatures. (From Buxton, 1931*a*.)

There are several ways of expressing humidity (see fig. 1), and of these the measure 'saturation deficit', which is the *amount* by which the water vapour present in a sample of air falls short of the saturation value, appeared to offer the best solution of the problem, for saturation deficit was known to represent the drying power of the air fairly closely and it formed a convenient means of expressing the combined effects of temperature and humidity in this respect.

2

This measure was therefore adopted by entomologists, and Buxton (1931b, 1932a) claimed that 'it appears that we have arrived at the physical law which governs the loss of water from an insect'. Mellanby (1935a), in a review of the literature on water loss, concluded that if insects in identical morphological and physiological states are exposed to different humidities, the rates at which they lose water will be proportional to the saturation deficit provided this is not above a maximum figure peculiar to the species; furthermore, if the rate of water loss from different individuals is not proportional to saturation deficit then their morphological and physiological states cannot be the same.

This generalization, known as the 'saturation deficit law', provided a useful stimulus, and a great deal of work was done to investigate its validity. Unfortunately much of this work, though valuable on other grounds, is not relevant to the question of the validity of the law, for although the law refers to loss of water, hardly any measurements of loss of water were made, and observations were in terms of longevity or of limiting conditions, on the assumption that a fixed depletion of water caused death.

Confusion also arose from the fact that saturation deficit, as a measure of drying power, may determine the rate of transpiration, but it is very unlikely to determine total water loss including excretion and defaecation.

Considerations of space forbid an examination of this field of work in detail. For further information the reader is referred to a valuable review by Johnson (1942), and to papers by Ludwig (1945) and Edney (1945, 1947).

THE PHYSICS OF EVAPORATION. A different approach to the problem was adopted by Ramsay (1935a) who investigated the physics of *evaporation* in relation to biological systems. Ramsay pointed out that even in inanimate systems the relation between temperature, humidity and evaporation is a very complicated one. When a small surface is evaporating into an infinitely large atmosphere unsaturated with water vapour, the concentration gradient may be considered as non-existent at a finite distance away, and a steady state will be reached in which

$$E = K\,(p_0 - p_d),$$

3

where $E =$ rate of evaporation of water,

  $p_0 =$ partial pressure of water vapour in air saturated at the temperature of the surface,

  $p_d =$ partial pressure of water vapour in air a short distance away from the surface,

  K is a constant proportional to T^r/P, where $T =$ absolute temperature, $P =$ total pressure, and r is a constant whose value is approximately 2.

Two important considerations arise. First, unless the temperature of an evaporating surface is the same as that of the air above it (and owing to the cooling effect of evaporation this is seldom true), $p_0 - p_d$ will not be equal to the saturation deficit of the air. Secondly, the factor K itself includes a factor depending upon the square (approximately) of the absolute temperature, so that even in an ideal system the rate of evaporation is not proportional to the saturation deficit, for a rise in temperature from 0° to 30° C. will increase evaporation by about 25 per cent even at the same saturation deficit.

In still air in a small closed container, further complications arise, for the rate of evaporation is then affected by the dimensions of the container. In practice, when arthropods are confined in a desiccator together with a humidity controlling substance, it is unsafe to assume that the animals are exposed to the humidity immediately above the controlling substance, for a gradient will be set up. The gradient is much steeper and shorter, and the results therefore more accurate, if moving air is used.

Ramsay (1935*b*) used moving air in which to measure the effects of humidity and temperature on transpiration from the cockroach. At constant temperature he found the rate of transpiration from the spiracles to be proportional to saturation deficit. When the temperature was raised from 25° to 30° C. transpiration increased by about 10 per cent, against an expected increase of 7 per cent. Evaporation from the general surface, too, was approximately proportional to saturation deficit, but only at any one temperature.

The saturation deficit law in its general form is therefore theoretically unsound. Nevertheless it may be an advantage to know that when adequate precautions have been taken, there

4

is no well-authenticated body of information conflicting with the assumption that water evaporates from arthropods at a rate proportional to the saturation deficit *provided the temperature is constant*, and there is some evidence that it does. There is also evidence, most of it indirect, that in some cases the effect of temperature alone on evaporation is slight, so that there is here a closer approximation to the general form of the law. It would not, however, be at all safe to assume that the law applies in unknown cases.

As we shall see below, recent work has shown the situation to be much more complex than was at first thought. The main reason for this is that temperature, and probably humidity, besides affecting the drying power of the air, also affect the permeability of the cuticle itself.

The third important physical factor affecting evaporation is wind speed, and we have far too little information of the effect of this upon animals. The faster the air is moving over an evaporating surface, the steeper and shorter the gradient of water-vapour concentration next to the surface, and therefore the greater the rate of evaporation. The precise relation between wind speed and evaporation is again very complex. On theoretical grounds it is to be expected that at low velocities evaporation from a free water surface will be proportional to the square root of the velocity, and at higher velocities, owing to turbulence, the rate will be linearly related to velocity. Ramsay (1935*a*) measured evaporation from a porous pot in a wind tunnel and found that above 4 m./sec. evaporation was linearly related to velocity. The effect of wind speed upon transpiration from a cockroach with the spiracles either open or occluded was also measured. There are too few determinations to say whether the rate is linearly related to velocity, but the effect of increasing the velocity was much greater on evaporation from the spiracles than from the general surface.

CHAPTER 2

TRANSPIRATION AND CUTICLE STRUCTURE

TRANSPIRATION FROM RESPIRATORY SURFACES. The diffusion constant of oxygen in chitin is only about one-thirtieth of that in water (Krogh, 1919), and unless the respiratory surfaces are moist, the supply of oxygen to the tissues would be insufficient. These surfaces are therefore a potential source of rapid transpiration, and it is not surprising to find that in well-adapted land arthropods they are tucked away within the body or covered by opercula. Loss from the respiratory surfaces nevertheless forms a significant proportion of the total water loss; without protection life in anything but saturated air would be impossible.

In the Onycophora there is a very large number of simple, unbranched tracheae which open by small unprotected spiracles, and the rate of water loss is very high (Manton and Ramsay, 1937; Morrison, 1946).

Respiration in the terrestrial isopods has been reviewed by Edney (1954). These animals possess pairs of biramous pleopods which function as respiratory organs (Remy, 1925; Mödlinger, 1931). They show varying degrees of adaptation to terrestrial existence (Meinertz, 1944) from the unmodified gill-like structures of the littoral genus *Ligia* to the specialized condition in *Porcellio* and *Armadillidium*, where the first two pairs of pleopodal exopodites bear internal tufts of tubules known as pseudotracheae opening to the exterior by a single pore without a closing device. Verhoeff (1920) thought that the gill surfaces were moistened by secretion from 'Weber's glands' at their base. But Gorvett (1950) showed that no such glands exist, and it now seems probable that the gills are moistened by diffusion from within, and sometimes by external water. Loss of water from the gills accounts for some 40 per cent of the total loss in

Porcellio scaber, and 20 per cent in *Ligia oceanica*, but transpiration per unit area from the gills is much greater in *Ligia* than in *Porcellio* (Edney, 1951a). Transpiration from the general surface of woodlice is rather rapid, and Edney and Spencer (1955) found that surfaces other than the pleopods are used for oxygen uptake. Provided the air is moist and the integument is not dry, *Ligia* can absorb sufficient oxygen for survival either through the pleopods or through the rest of the integument. The faculty is present to a decreasing extent in the more terrestrial forms. Total oxygen uptake falls off in dry air, and there is evidence that in prolonged exposures to dry air, death is due in part to oxygen lack. The respiratory organs of isopods are therefore still unsatisfactory for really dry habitats.

In spiders the main respiratory organs are a pair of book lungs, situated in cavities within the abdomen, and provided with well-developed closing mechanisms (Kastner, 1929), controlled by carbon dioxide tension in the tissues (Hazelhoff, 1926; Jordaan, 1927). As regards water loss through the spiracles, Davies and Edney (1952) found that if the spiracles of *Lycosa amentata* are kept open by carbon dioxide, loss of water is increased by about 50 per cent above normal.

Scorpions, too, possess book lungs, arranged in four pairs on the ventral surface of the mesosome. De Buisson (1928) described the closing mechanism of their stigmata, and Fraenkel (1930) observed a ventilation mechanism involving both cardiac and pulmonary movement. Zoond (1931, 1934) found that blocking all the stigmata of *Opisthophthalmus* completely prevented oxygen uptake and soon led to the death of the animal, but if only one stigma was left open, this provided sufficient oxygen for normal needs. Millot and Paulian (1943) confirmed these results in *Buthus australis*. Zoond found no cutaneous respiration whatever in *Opisthophthalmus*. This is a little surprising, but may be accounted for by an extremely dry cuticle, for the rate of transpiration from scorpions is remarkably low for a cryptozoic animal (Cloudsley-Thompson, unpublished). There is no published information on loss of water from the book lungs, but Cloudsley-Thompson found little or no effect upon total loss when they were blocked.

Ticks possess a tracheal system similar to that of insects, and the spiracles are provided with closing mechanisms (Mellanby,

7

1935*b*; Browning, 1954*a*) similar in function to those of insects. Mellanby found an increase in water loss when the spiracles were kept open, but the situation is complicated by the fact that many ticks are capable of absorbing water through the cuticle in moist air (Lees, 1946*a*, 1947).

The tracheal system of diplopods consists of unbranched tubules opening by large numbers of tracheal pockets without closing mechanisms (Miley, 1930). Chilopods, however, possess a well-developed tracheal system (Verhoeff, 1941) with protected spiracles.

If the spiracles of insects are kept open either as a result of metabolism or by exposing them to carbon dioxide, the rate of transpiration increases greatly. Ramsay (1935*b*) found that this increased from 3·9 to 6·0 mg./hr. in stimulated *Blatta*, and Mellanby (1934*b*) found that in *Xenopsylla* transpiration was doubled in 5 per cent carbon dioxide (fig. 2). The mealworm, even after starving in dry air for four months, lost water through the spiracles when these were kept open by carbon dioxide as rapidly as it did after only two weeks' starvation. During this time, metabolism was reduced and the total rate of water loss fell owing to the spiracles being open for a shorter proportion of the time (fig. 2). Similarly, *Rhodnius*, which is normally resistant to desiccation, died in three days in dry air if the spiracles were kept open (Wigglesworth and Gillett, 1936).

Nevertheless, even in normally respiring insects a large proportion of the water lost passes through the spiracles: 60 per cent in pupae of *Bombyx mori*, and 70 per cent in *Gastrimargus* adults (Koidsumi, 1935). But here the absolute total loss is very low, and the high proportional loss through the spiracles is testimony to the effective water-proofing of the rest of the integument.

In some small arthropods, such as podurids, there is no tracheal or other internal respiratory system, and gaseous exchange is entirely cutaneous. W. Davies (1928) showed that of several species of Collembola, *Sminthurus viridis*, the only one with tracheae, was the most resistant to dry conditions. Cutaneous respiration can only be effective in small animals living permanently in a moist environment, but there is evidence of a certain amount of cutaneous absorption of oxygen even in fully terrestrial forms: thus Buddenbrock and Rohr (1922) found that up

8

to 25 per cent of the total oxygen uptake in *Dixippus morosus* was cutaneous, and Fraenkel and Herford (1938) demonstrated cutaneous uptake amounting to some 10 per cent of the total in *Calliphora* larvae, in *Chaerocampa* (Lep.) larvae, and in *Tenebrio* larvae. It is probable that such cutaneous respiration occurs to a greater or less extent throughout the arthropods (except perhaps in scorpions), but it is clear that no animal which is well adapted to terrestrial conditions can afford to use the general body surface as the chief means of oxygen absorption, and the evolution of an efficient internal respiratory system, together with effective spiracular control, has been a *sine qua non* for full exploitation of the terrestrial environment.

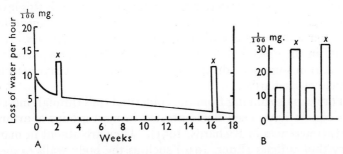

FIG. 2. The effect of opening the spiracles on the rate of water loss. A, *Tenebrio* larvae; B, *Xenopsylla* adults. At x the spiracles were kept open with 5% carbon dioxide. (From Wigglesworth, 1950, after Mellanby, 1934b.)

GENERAL ORGANIZATION OF THE CUTICLE. This subject has been reviewed so far as insects are concerned by Wigglesworth (1948a), and more recently for arthropods as a whole by Richards (1951). It is a rapidly expanding subject, and important new developments on certain aspects have recently come to light.

The cuticle of arthropods is secreted by a single-layered epidermis and consists of two fundamental layers, an outer, thin epicuticle, and an inner, usually much thicker, endocuticle (fig. 3). The essential difference between these layers is the absence of chitin in the former. The endocuticle (procuticle of Richards) consists essentially of a protein and chitin matrix; it may remain apparently unchanged, or the outer regions may become hardened and darkened (sclerotized) by tanning, to give an outer exocuticle, the inner unchanged portion still being

9

referred to as endocuticle; or the outer portion may become calcified as well as being tanned, as in many Crustacea.

The epicuticle itself is very complex; its nature probably varies from one group of arthropods to another, and there is still a good deal of doubt concerning its minute structure in all groups. Most epicuticles appear to consist of at least two layers (Richards and Anderson, 1942), an inner proteinaceous and an outer lipoid. The inner layer apparently consists, at least in most insects, of a lipoprotein complex which may be tanned by polyphenols as in the exocuticle. For this material the term 'cuticulin' (Wigglesworth, 1933a) has been generally accepted. The outer lipoid layer, which may in fact not be a continuous layer, is composed of wax or grease in insects, and it is this layer which is believed to confer the large degree of impermeability to water found in the cuticles of many terrestrial species. Also found in the epicuticles of many arthropods are polyphenols— perhaps in some as a distinct layer, and a cement layer (tecto-cuticle of Richards).

Pore canals running from the epidermal cells through most of the cuticle have been demonstrated in most groups of arthropods (references in Richards, 1951). They are absent in many very thin cuticles (Eder, 1940) such as the body walls of mosquito larvae and all arthropod tracheae, and also apparently from the thick inner endocuticle of *Sarcophaga* larvae (Dennell, 1946). In some cases the pore canals undoubtedly penetrate the cuticulin layer of the epicuticle, as in *Rhodnius* (Wigglesworth, 1947) but not in *Periplaneta* as shown by the electron microscope (Richards and Anderson, 1942). The cuticle is by no means homogeneous over the whole animal, and it may be interrupted in many ways. It is thinner, and the hardened exocuticle is absent at the intersegmental membranes, and it is interrupted completely by the ducts of the dermal glands when these are present. But the epicuticle is universally present except perhaps over the sense organs (Slifer, 1954, 1955).

Permeability to water of the arthropod cuticle varies enormously. To some degree, impermeability may be conferred by sclerotization (Lafon, 1943; Kalmus, 1941). Nevertheless, there is no general correlation between impermeability and darkness or rigidity of the cuticle, for soft-skinned insects such as *Tineola* larvae may be as resistant to desiccation as heavily armoured ones (Mellanby, 1934a), and there is conclusive evidence that the property of impermeability is conferred by the epicuticular waxes.

THE EPICUTICLE OF INSECTS. Evidence regarding the water-proofing properties of epicuticular waxes was first derived from studies on the formation of the new epicuticle after a moult. In *Rhodnius prolixus* the process was studied by Wigglesworth (1947). The first part to be laid down is the inner proteinaceous layer, which immediately after formation gives a

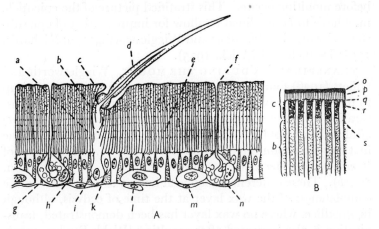

FIG. 3. A, ideal section of the insect integument; B, schematic section of the epicuticle: *a*, laminated endocuticle; *b*, exocuticle; *c*, epicuticle; *d*, bristle; *e*, pore canals; *f*, duct of dermal gland; *g*, basement membrane; *h*, epidermal (hypodermal) cell; *i*, trichogen cell; *k*, tormogen cell; *l*, oenocyte; *m*, haemocyte adherent to basement membrane; *n*, dermal gland; *o*, cement layer of eipcuticle; *p*, wax layer; *q*, 'polyphenol layer'; *r*, cuticulin layer; *s*, pore canal. Dennell and Malek (1955) find in *Periplaneta* a paraffin layer in place of Wigglesworth's polyphenol layer. (From Wigglesworth, 1948a.)

positive Millon's reaction, and is positive to Sudan black. Later this layer is negative to fat stains, but on warming in nitric acid and potassium chlorate it breaks down to give oily droplets. It is therefore thought to consist of lipoprotein, but neither the lipoid nor the protein has been identified. A few hours before moulting the surface of the epicuticle is argentaffin. Whether this is due to the presence of a polyphenol as Wigglesworth believed, or to some other argentaffin substance (Blower, 1951; Dennell and Malek, 1955), is uncertain. What is important is that immediately before moulting, the surface is no longer

11

argentaffin unless treated with cold chloroform and this suggests the presence of a lipoid layer. Within an hour or so after the moult, cold chloroform fails to reveal the argentaffin layer, because a cement (tectocuticle of Richards) has been secreted over the lipoid layer.

In view of these findings Wigglesworth proposed as a working hypothesis a four-layered structure for the epicuticle of *Rhodnius* (fig. 3*b*): cuticulin, polyphenol, wax and cement; the cuticulin layer being tanned by oxidation products of the polyphenols before moulting occurs. This stratified picture of the epicuticle may have to be modified to allow for impregnation of one substance by another (Kramer and Wigglesworth, 1950; Richards, 1951; Dennell and Malek, 1955).

TRANSPIRATION DURING THE MOULT. Wigglesworth and Gillett (1936) found that in *Rhodnius* no extra water was lost up to the time of ecdysis, but that after the moult the rate of transpiration was more than doubled for a few hours. In *Tenebrio* adults, the rate of loss is from four to six times normal during the first day after moulting (Wigglesworth, 1948*b*). In the larva of *Tenebrio* moulting does not affect water loss at all (Buxton, 1931*b*). These differences may well be due to differences in the completeness of the wax layer at the time of ecdysis, although in woodlice, where no wax layer has been demonstrated, transpiration is also increased after moulting (Webb-Fowler, 1955). It is significant that nearly all the old endo- and exocuticle of *Rhodnius* (and of most other insects) is digested before the skin is cast, leaving only the thin epicuticle. Yet if the old skin is removed at that stage, the new cuticle is still hydrophilic and very permeable to water. Water-proofing must therefore reside in the old epicuticle. Just before the old skin is due to be cast, permeability of the new cuticle is rapidly reduced, and this coincides in time with the protection of the argentaffin layer by a chloroform-soluble material. The conclusion must be, therefore, that this outer layer of the cuticle is responsible for water-proofing.

In *Tenebrio*, the same methods of examination reveal the same process of formation and structure of the epicuticle (Wigglesworth, 1948*b*). In the blow-fly larva, *Sarcophaga*, however, Dennell (1946) recognized only two constituent layers, neither penetrated by pore canals.

Comparative work on the cuticles of other arthropods (references in Richards, 1951) shows a reasonably consistent structure throughout the group. Part of the procuticle is usually tanned and sometimes subsequently impregnated with lipoids. It may also be hardened in other ways. An epicuticle is probably universally present, characterized by the absence of chitin, but usually containing lipoid material which in insects and ticks at least is laid down as a lipoprotein complex and subsequently tanned. Only in insects and arachnomorphs has the existence of a continuous, discrete wax layer been proposed.

As mentioned above, the picture of a discrete homogeneous wax layer above the cuticulin and polyphenol layers may have to be modified. In *Rhodnius* and in other insects (Wigglesworth, 1945) and in the tick *Ornithodorus* (Lees, 1947), if the cement and wax layers are damaged, fresh wax may be secreted by the epidermal cells through the cuticle. This may be imagined to occur through the pore canals, but Kramer and Wigglesworth (1950) suggest that grease secretion in the cockroach *Periplaneta* may be continuous throughout the life of the insect and occur through the cement layer, which is more difficult to understand. Again, in the egg of *Ornithodorus*, Lees and Beament (1948) have some evidence that the water-proofing wax gradually permeates the shell from the surface, and the same is true of the serosal membranes in the egg of *Rhodnius* (Beament, 1949). For these and other reasons to be discussed below (p. 90) it may be necessary, as Kramer and Wigglesworth point out, to begin to think of the various layers as being penetrated to a greater or less extent with diffusing waxes. The polyphenol 'layer' has also been called in question by Richards (1951), who points out that polyphenols are also known to be present in the sub-layers, and by Dennell and Malek (1955).

Nevertheless there is evidence that in some insects and ticks the important water-proofing layer is at or near the surface, and to this we may now turn.

THE EFFECT OF ABRASIVE DUSTS. It was found by Alexander *et al.* (1944) that insects exposed to inert dusts, such as alumina, died as a result of water loss, and in general the smaller the size of particle the greater the effect (Alexander *et al.*, 1944; G. Jones, 1955). Wigglesworth (1944, 1945) and Lees (1946a) showed that the effect of the dusts was to abrade the surface of

13

the epicuticle only (revealing argentaffin properties below) and to increase the rate of transpiration several times. The distribution of scratches revealed by the argentaffin test showed that abrasion occurred only between moving parts in contact, or where the abrasive was actually rubbed on the insect. Dead insects sprinkled with dust, or living insects sprinkled and then prevented from moving showed no increase in water loss. In the cockroach, however, where the surface lipoid is a mobile grease, sprinkling without rubbing was effective in increasing transpiration, and in this case it was thought that adsorption of the grease on to the dust particles had occurred. Beament (1945) found that certain dusts could adsorb insect waxes deposited on gelatin membranes, but not when the film was deposited on an insect epicuticle—abrasion was then necessary. The difference he ascribed to differences in orientational force upon the wax molecules exerted by the two kinds of membrane.

Hurst (1948), on the other hand, claimed that the action of dusts was due to their adsorptive properties only, and that dusts were effective in increasing transpiration from dead insects. G. Jones (1955) also reported that dusts increase transpiration even from dead and motionless bees, and concluded that they exert their effect by absorption. Holdgate (1955a), however, believes that the possibility of abrasion was not entirely excluded in Jones's work.

It is not impossible that both adsorption and abrasion are involved, though to different extents in different insects, depending upon the thickness of the lipoid layer, its consistency and its distribution. But whatever the mechanism, the results of this line of inquiry provide additional evidence that the main water-proofing layer of the epicuticle is at or near the surface.

TEMPERATURE AND PERMEABILITY. Gunn (1933) was the first to measure the effect of temperature on transpiration from an insect. He used the cockroach *Blatta orientalis* and found that both in dry air and at 50 per cent R.H. transpiration increased more rapidly with temperature above 30° C. than below that figure, even after allowing for the rise in saturation deficit at higher temperatures (fig. 4). Gunn attributed the rise in transpiration rate to a change in respiratory mechanism. Active pumping of air in and out of the spiracles, he observed, commenced above about 30° C. But Ramsay (1935b), working with *Periplaneta,* found that a similar increase in transpiration occurred even in dead animals with the spiracles blocked. This focused attention upon the cuticle itself, and opened up a new line of approach to problems of water loss and cuticle structure in arthropods.

Ramsay's insects were exposed in air at 50 per cent R.H. at temperatures from 20 to 50° C. The results he obtained are

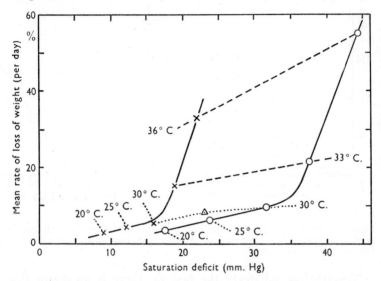

FIG. 4. Rate of loss of weight of the cockroach at various temperatures and humidities. O=dry air; △=26% R.H.; ×=50% R.H. (From Gunn, 1933.)

represented in fig. 5*a.* Recognizing the need to allow for increase in saturation deficit and therefore of drying power of the air at higher temperatures, he extrapolated from the values at each temperature (on the assumption that the rate of evaporation was proportional to saturation deficit at any one temperature) and found the value of each extrapolation at a line representing a saturation deficit of 36 mm. mercury. These values he then graphed (fig. 5*b*) and obtained a curve expressing the relation between rate of evaporation and temperature at one saturation deficit.

Now the points at the lower temperatures on this curve are imaginary, for 32° C. is the lowest temperature at which air can have a saturation deficit as great as 36 mm. mercury.

This does not matter so much, for in theory at least a saturation deficit of 10 mm. mercury could have been chosen, and the relative distance between the points representing different temperatures would presumably have been the same. But there seems to be no reason why two straight lines rather than one continuous curve should be drawn through the points in fig. 5*b*.

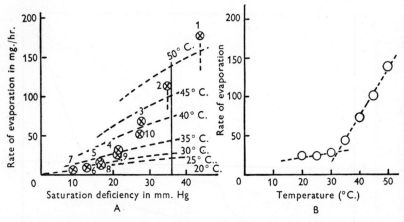

FIG. 5. Effect of temperature on the rate of evaporation from the body surface of the cockroach. B is derived from A when saturation deficiency is 36 mm. Hg. (From Ramsay, 1935*b*.)

Nevertheless, whatever the best fit, there is no doubt that permeability is increased with rising temperature, and Ramsay suggested that this might be due to a phase change, possibly melting, of a fatty substance in the cuticle. He does not appear to have contemplated a continuous change with temperature of the permeability of the fatty substance, yet this is what his figures suggest. Consider a system in which water evaporates through a membrane into an atmosphere of constant saturation deficit, then if due allowance is made for increased rate of diffusion at higher temperatures, the graph of rate of evaporation upon temperature will be a horizontal straight line. If at a given temperature a sudden increase in permeability occurs, this will have the effect of putting a sharp step in the curve, which will continue at higher temperatures as a horizontal straight line at a higher level. A continuous rise in the curve, on the other hand, must mean a continuously increasing permeability of the membrane, provided that permeability is the limiting factor. Ramsay's graph seems to have been interpreted by subsequent workers as indicating a sudden, discontinuous change in permeability.

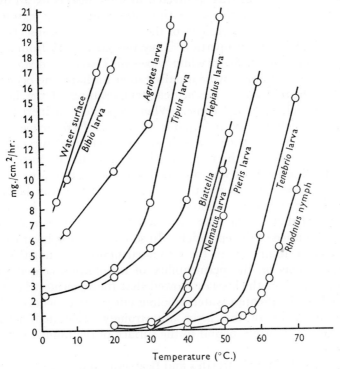

Fig. 6. The rate of evaporation of water from dead insects at different temperatures. (From Wigglesworth, 1945.)

It was not until ten years later that the next important step was taken, when Wigglesworth (1945) measured the rate of transpiration at various temperatures in a variety of insects, and found that apart from aquatic and burrowing insects, transpiration increased rather abruptly above a certain temperature which varied from one species to another (fig. 6). This was called the 'critical temperature'. In the same paper Wigglesworth reported the effects of superficial abrasion described above, and of the action of wax solvents and detergents upon the cuticle, all of which had the effect of increasing permeability greatly and of abolishing the critical temperature. These facts led to the conclusion that the lipoid material already known to be present in the insect cuticle (Wigglesworth, 1933a; Bergmann, 1938; Pryor, 1940; Hurst, 1940) was a superficial layer of wax, protected in some insects by an overlying cement. At the same time Beament (1945) extracted waxes from the cast cuticles of the same species of insects and investigated their properties. They proved to vary from a soft grease in *Blatta* to hard white waxes in *Tenebrio* and *Rhodnius*. Their melting points were usually rather unsharp and occurred a few degrees above the critical temperatures of the insects from which they had come. When deposited on inert membranes (gelatin, or thoroughly de-waxed *Pieris* wings), the extracted waxes gave transpiration/temperature curves very similar to those of the respective intact insects. By measuring the total amount of wax obtained and the surface area of the insects concerned, Beament concluded that the wax layer in the living insect was about 0·25 μ thick. There is, however, no proof that all the wax extracted came from a single layer at the surface of the epicuticle. In so far as such a layer exists, it may be thinner than the estimated figure, but not thicker. Experiments on the permeability of membranes coated with wax of various thicknesses showed that the innermost layer of wax had the greatest water-proofing effect. It is believed that the molecules of this layer are strongly orientated, tightly packed, and firmly bound to the surface of the tanned protein of the epicuticle.

Now both Wigglesworth's and Beament's data were obtained in dry air, and no allowance was made for saturation deficit or diffusion effects. It would be incorrect, therefore, to deduce the effects of temperature alone from the shapes of these curves until they are corrected. Similarly shaped curves have been found for *Diataraxia* (Way, 1950), ticks (Lees, 1946a, 1947), spiders (Davies and Edney, 1952) and scorpions (Cloudsley-Thompson, unpublished). These again were obtained in dry air and the same comment applies.

Recently, however, Mead-Briggs (1954), in my laboratory, and Holdgate (1955a) have independently re-examined the problem. Mead-Briggs exposed insects to a slowly moving stream either of dry air or of air whose relative humidity was so adjusted that at temperatures from 30 to 60° C. the saturation deficit was constant at 30 mm. mercury. In the second case, the drying power of the air at each temperature was demonstrably the same, for evaporation from a free water

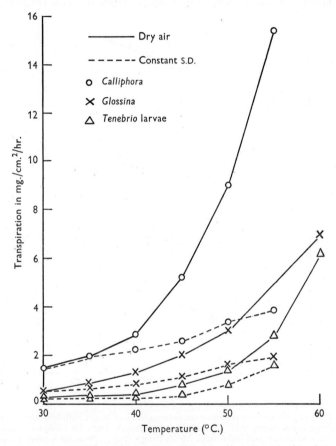

Fig. 7. The relation between transpiration and temperature in dry air and in air at constant saturation deficit. (Data from Mead-Briggs, 1954.)

surface exposed in the same apparatus was constant. His results for *Tenebrio* larvae, *Glossina* and *Calliphora* show that at constant saturation deficit, and after allowing for the increased rate of diffusion at higher temperatures, there is still a residual rise in transpiration rate with temperature (fig. 7), and this

19

may be ascribed to the effect of temperature on permeability of the cuticle. The rise, however, is a gradual, continuous one, with no sign of a break in the curve.

Holdgate used dry still air, and applied a factor to his results to allow for saturation deficit and diffusion. His results also

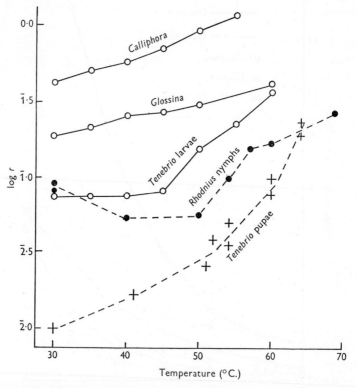

Fig. 8. The effect of temperature on permeability of insect cuticles; $r = 10^6 \times$ rate of transpiration in mg./cm.²/hr./mm.Hg/abs. temp.[2] (Data for *Calliphora*, *Glossina* and *Tenebrio* larvae (full lines) from Mead-Briggs (1954), those for mature *Rhodnius* nymphs and mature *Tenebrio* pupae (broken lines) from Holdgate (1955a).)

show that, in *Rhodnius* and *Tenebrio* larvae, there is a large residual temperature effect. When these results of Holdgate and Mead-Briggs are graphed as log (transpiration rate) on temperature (fig. 8) the resulting curve is approximately linear for *Calliphora* and *Glossina*, but cusped for *Tenebrio* larvae, *Tenebrio* pupae and *Rhodnius*. In other words, the transpiration rate is approximately exponentially related to temperature, perhaps with the addition of some other factor or factors in *Rhodnius* and *Tenebrio*.

If the figures obtained by Mead-Briggs using dry air are corrected for saturation deficit and diffusion, these corrected results should theoretically lead to similar values to those he obtained in constant saturation deficit. In fact (Table 1) they are considerably lower in the case of *Tenebrio* larvae and *Glossina*, and somewhat lower in *Calliphora*. This may be the result of some factor other than the drying power of the air limiting evaporation at higher temperatures in dry air; if evaporation is rapid, insects may dry up and consequently lose water less rapidly, as Bursell (1955) found for woodlice (p. 22).

TABLE 1. *Rates of transpiration determined either in dry air or at a constant saturation deficit of 30 mm. Hg, and reduced to $\mu g./cm.^2/hr./mm.$ Hg saturation deficit*

(Data from Mead-Briggs (1954, and unpublished work).)

	35° C.		45° C.		55° C.	
	Dry air	s.d. 30	Dry air	s.d. 30	Dry air	s.d. 30
Tenebrio larvae	5·5	6·0	7·6	10·8	20·3	50·9
Calliphora	6·1	6·3	6·5	8·3	11·2	12·7
					(60° C.)	
Glossina	2·1	2·0	2·5	3·7	3·9	5·6

These results will be referred to again, meanwhile we may consider the effects of temperature on evaporation from other groups of arthropods.

TRANSPIRATION IN OTHER ARTHROPODS. Ticks (Lees, 1947), spiders (Davies and Edney, 1952) and scorpions (Cloudsley-Thompson, unpublished) show the same-shaped transpiration/temperature curves as insects do in dry air.

In woodlice, Edney (1949, 1951a) found that transpiration is a good deal higher per unit area than in most insects and that even in dry air there is no transition or critical temperature, for the shape of the transpiration/temperature curves was much the same as that of the saturation-deficit/temperature curve for dry air (fig. 9). Furthermore, when *Porcellio* was exposed to constant saturation deficits at 5° C. intervals from

10 to 45° C., transpiration was approximately constant after allowing for diffusion effects. Mead-Briggs (1954) found the same for *Oniscus*.

Auzou (1953) and Bursell (1955) have also measured transpiration from woodlice, and they both find that the rate falls off, sharply at first, with time. The slower decrease which

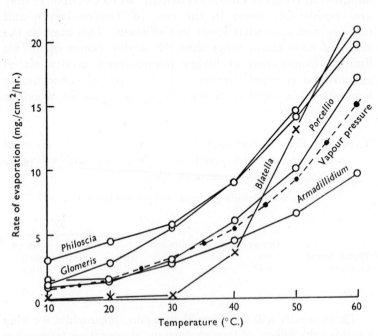

FIG. 9. The rate of evaporation of water from various woodlice, the millipede *Glomeris* and a cockroach (*Blatella*) at temperatures from 10° to 60° C. in dry air; the animals exposed for 1 hr. periods (From Edney, 1951*a*.)

occurs later is caused, according to Bursell, by a change in permeability of the cuticle brought about by an increase in electrolyte concentration in the haemolymph as water is lost.

Bursell proposes two lipoids in the endocuticle of *Oniscus* to account for two breaks in the transpiration/temperature curve, but his figures do not differ very conspicuously from a saturation deficit curve if allowance is made for diffusion. Auzou, on the other hand, claimed to have demonstrated a 'critical temperature' in *Porcellio*, though she did not find one in

Oniscus. Her contention, however, is based on one point only, and that of unknown validity, in the transpiration/temperature curve. In view of these apparent contradictions it would probably be unwise to build up pictures of cuticle structure based on transpiration rates before the possible effects of differences in technique have been eliminated.

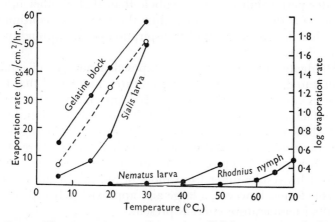

Fig. 10. The evaporation rate of water from *Sialis* larvae compared with some terrestrial insects. ○ are points representing the logarithm of the evaporation rate for *Sialis* larvae at the same temperatures. (From Shaw, 1955a.)

As regards the diplopods, Cloudsley-Thompson (1950) working with the millipede *Paradesmus* and Edney (1949, 1951a) with *Glomeris*, found a linear relation between saturation deficit and transpiration, with no sign of a transition temperature. The same is true of the centipede *Scolopendra* (Cloudsley-Thompson, unpublished).

TRANSPIRATION IN AQUATIC INSECTS. Shaw (1955a) investigated permeability of the cuticle of *Sialis* and found this to be much greater than in terrestrial insects (fig. 10). When exposed to dry air he found the usual rise in transpiration rate with temperature, but with no transition temperature. The rise was much greater than would be accounted for by saturation deficit and diffusion, and there is a fairly good exponential relationship between his uncorrected figures and temperature. (This is different, of course, from the situation discussed above in terrestrial insects, where, *after* correction for saturation deficit and diffusion, transpiration was found to be exponentially related to temperature.)

Shaw proposes a continuous wax layer to account for the transpiration curve, and he succeeded in isolating a wax from the cuticle, which, if it were in the form of a continuous superficial layer, would be 0·1 μ thick. But in view of the much greater permeability of *Sialis* cuticle than that of other insects, there is a distinct possibility that the wax is discontinuous, and that the greater part of transpiration occurs through pores in the wax.

Holdgate (1955*a*, 1956) also exposed a number of aquatic insects to dry air at various temperatures, and found that their cuticles vary considerably as regards permeability. After applying a factor to allow for saturation deficit and diffusion, his results show that in the more permeable forms (including *Sialis*) transpiration is high and shows no residual temperature effects (in other words, it is proportional to saturation deficit). In this respect his results differ from those of Shaw. In the less permeable forms, however, a large and approximately exponential residual temperature effect is present.

INTERPRETATION OF TRANSPIRATION/TEMPERATURE CURVES. It may be useful at this stage to summarize the evidence on the relation between transpiration and temperature, and to relate this, so far as possible, to cuticle structure. This is clearly a rapidly developing subject, and work recently published or still in progress makes it likely that many of our ideas will have to be reconsidered. Any conclusions drawn now must therefore be regarded as tentative: they will certainly have to be modified in the light of fresh evidence.

It is unfortunately all too clear that contradictions exist in the information already available; thus Auzou, Bursell and Edney all obtained somewhat different results for woodlice; and it is obviously impossible to allow for the case of *Sialis* larvae, where Holdgate (1955*a*) finds transpiration to be entirely accountable for in terms of saturation deficit and diffusion, while Shaw (1955*a*) finds a residual temperature effect. Some of these differences may be ascribed to differences in technique, and another contributory factor is that, with few exceptions, authors (including the present writer) have not been at pains to establish the validity of the points on their graphs (indeed, some of the published curves show no points at all). In other words, we do not know the variance of these points, and therefore have no means of judging the probability of the curves drawn through them being correct.

A glance at the variety of curves obtained by different authors using different techniques shows that any hypothesis designed to fit them all would have to be immensely complicated. Some simplification at this stage is therefore both necessary and legitimate.

In general terms, then, the situation to be considered is as follows: in arthropods such as woodlice, millipedes and aquatic insects with a high rate of transpiration, there is little or no effect of temperature upon the permeability of the cuticle. In those with less permeable integuments, such as most terrestrial insects, spiders and ticks, and perhaps some aquatic forms, there is an effect of temperature upon permeability which appears to be greater where the permeability is less. In these animals, the relation between permeability and temperature is approximately exponential. In no case is there a clearly marked 'critical' or 'transition' temperature, although there may be a cusp in the otherwise linear curves of log (transpiration)/temperature for low-permeability cuticles.

Now there appear to be several processes which could lead to an increase of permeability with temperature. Firstly, since the wax is now known to consist of a number of different components (p. 26), it is possible that phase changes in these components occur over a range of temperatures, and lead to an apparently continuous rise of permeability with temperature. Secondly, as Barrer (1939, 1941) and Doty et al. (1944) have shown, there are reasons why high-polymer membranes must be expected to show an exponential rise of permeability with temperature. The steepness of the rise depends upon the shape and size of micropores in the membrane, and according to Barrer's work is greater in intrinsically less permeable membranes.

Both these possibilities refer to a situation where the wax is considered as substantially continuous (save for micropores). But if the wax is discontinuous, a third possibility exists, namely that transpiration occurs at least partly through spaces, permeability being temperature-dependent as a result of

25

properties of the non-waxed cuticle. However, since removal of the wax, or abrasion, abolishes the exponential rise with temperature in those species which normally show it, it seems likely that transpiration through the non-waxed cuticle is essentially similar to evaporation from a free water surface (i.e. approximately proportional to saturation deficit) and that the properties conferring exponential temperature-dependence reside in the wax.

It is possible that either or both of the first and second processes mentioned above operate. What is important is that we cannot at present discriminate between their effects. It is possible also that in some arthropods, transpiration takes place partly through interruptions in the wax and partly (at higher temperatures) through the wax itself. If this is so, a variety of curves such as those obtained might be expected.

Perhaps the simplest hypothesis at present is that where there is no wax (woodlice, myriapods, and possibly very permeable aquatic insects), or where there are large pores in the wax (less permeable aquatic insects, and possibly more permeable terrestrial insects), evaporation occurs largely through a non-waxed surface and is not subject to an exponential temperature effect on permeability. But where the wax is more nearly continuous, and permeability lower (fully terrestrial insects, ticks and spiders), evaporation occurs partly through non-waxed pores and partly through wax, the latter component being subject to an exponential temperature effect which, rising rapidly at higher temperatures, sums with the other component to give a variety of *approximately* exponential curves, the precise shape depending on the degree of porosity.

CUTICULAR WAXES AND THEIR SOLVENTS. Further light has recently been thrown on the nature of cuticular lipoids by Beament (1955), who found that the cockroach secretes several members of the alcohol and paraffin series of compounds extending continuously from the longer-chain region, where they are hard, non-volatile waxes, into the shorter-chain region (about carbon 8 to carbon 12) where they are volatile and act as solvents for the waxes. This explains an observation by Wigglesworth that cockroach grease, if stored in air, changes to a hard, white wax similar to the wax of *Tenebrio* or *Rhodnius*. Mixtures of 8- and 10-carbon paraffins and alcohols proved to be miscible in all proportions with beeswax and with hard cockroach wax, forming greases which resemble the natural grease of the cockroach in possessing remarkable spreading power on water-saturated surfaces.

The vapour of an octane and octyl alcohol mixture decreases the permeability to water of artificially waxed membranes deposited from chloroform, but has no effect upon natural insect cuticles in this respect, while all other wax solvents increase the permeability. It would seem, therefore, that octane-octyl alcohol vapour encourages orientation of the long wax molecules at the water surface, and that chloroform and benzene vapours lead to disorientation, but there is no evidence that variation in the *amount* of solvent present affects the permeability of the natural membrane.

To what extent these solvents are present in other insect waxes remains to be seen. In the cockroach, continuous secretion of solvents must occur, perhaps by the oenocytes (Kramer and Wigglesworth, 1950), for the grease remains soft and evaporation of solvents from the surface is continuous (Beament, 1955). In other insects the wax is hard, but there is evidence that when secreted it is softer and more permeable. Thus *Rhodnius* and *Tenebrio* lose more water by transpiration in the first few hours following a moult than normally, and Holdgate (1955a) reports that permeability of *Rhodnius* cuticle increases at a lower temperature during the first few days after a moult than it does later.

The biological importance of the solvent constituents is great, for it secures an even distribution, and more important, a close orientation, of the wax molecules over the surface of the epicuticle. It is well known (Ramsay, 1935b) that cockroach grease will spread rapidly over water. Hard insect waxes cannot do this. It is also established that after abrasion, insects and ticks can repair the wax layer (Lees, 1947). Since the hard wax does not flow readily over the surface of the cuticle, the advantage of a spreading agent is obvious. Indeed, as Beament has remarked, the presence of a solvent makes it easier to accept the transport of waxes through the hardened cuticle during the process of repair.

In the same paper Beament (1955) showed that the cement, which in cockroaches is not superficial but lies within the grease, is very similar in properties to shellac, which is a surface secretion of the coccid, *Laccifer lacca*. Shellac is a complex association of a carbohydrate (laccose) with a small quantity of wax. Beament believes that what has previously been termed cement in many insects is in fact lac. The action of a boiling solvent in removing the wax from a cement-protected epicuticle would thus be a mechanical one, and this accounts for the absence of the cement layer as a thin lamina after such treatment.

The full implications of this work are not yet apparent. It will clearly be of great importance to know whether lac contributes to the water-proofing properties of the cuticle, and whether in some insects it forms a continuous layer. Meanwhile Beament's evidence regarding the properties of waxes and their solvents is certainly consistent with the concept of a continuous discrete layer if the action of the solvents is thought of as orientating the interface molecules upon a water-saturated surface; but the facts could equally well be interpreted if the solvent is considered to facilitate penetration into such a surface.

THE WETTING OF CUTICLES BY WATER. The surface properties of cuticles as regards wetting are clearly very relevant to the present discussion. There is much information of a descriptive kind arising from the need to know how insecticides behave when in contact with cuticles, but very little analytical work on the problem has been published.

Pal (1950) and Holdgate (1955*b*) have recently investigated the subject. Holdgate found that the contact angle between water and insect cuticles varies from 0° to 180°, and advancing angles are always a good deal larger than retreating angles. But since the roughness of a surface increases the angle if it is above 90°, and decreases it if it is below, and since air entrapped in the depressions of a rough surface may lead to very great increases in contact angle, the variation in true contact angle, as determined by the chemical composition of the surface, need be very much less.

It is an obvious advantage for a truly terrestrial insect to have a high contact angle which prevents wetting by rain and other water, and this is usually found to be the case. Often the natural angle is increased by hairs which entrap air.

Conversely, aquatic insects often have very low contact angles, and those such as pond skaters and *Gyrinus* which live on the surface of water, develop hair piles or structural modifications such that a constant water level on the body results.

Holdgate has confirmed for insect cuticles what was known for several physical materials, that contact angles are diminished

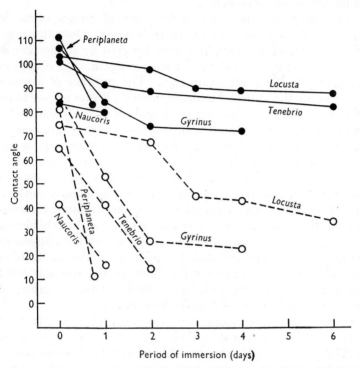

FIG. 11. Changes in the contact angles of several species following immersion in water. ●, — — : advancing angles; ○, — — : retreating angles. (From Holdgate, 1955*b*.)

markedly by prolonged immersion in water (fig. 11), so that the low angles found in aquatic insects may be a result of their environment and are not necessarily due to chemical differences in their cuticles. This explains the observation that *Aphelocheirus* larvae are very hydrophobic when newly moulted, but after immersion for a long period the contact angle falls to 10–20° (Crisp and Thorpe, 1948). Some aquatic insects,

29

however, maintain high contact angles throughout life; but the angles decrease after soaking when the insect is dead. This suggests the continuous secretion of material capable of maintaining orientation of the surface molecules, and Holdgate believes that the substances may be the short-chain alcohols and paraffins which Beament (1955) has shown to be solvents in the grease of cockroaches.

The most striking cases of lowering of angles occur in thoroughly dried cuticles and also in beeswax. Once the angle of cockroach-wing cuticle has been lowered by immersion in water, drying in air does not increase it again. But if kept in saturated air the angle again rises. The presence of a water phase inside the cuticle, Holdgate believes, is thus shown to be of importance in facilitating reorientation of the polar molecules, and this corresponds with Beament's (1955) observation that the presence of water in a membrane is necessary for the formation of a well-orientated film of cockroach grease.

Maintenance of a high contact angle is of importance in insects which breathe at the surface, and particularly in those like *Aphelocheirus* which breathe below the surface by means of a plastron. In *Aphelocheirus* retention of a gas film is achieved by the fine pile of hairs which results in a very high apparent contact angle (Thorpe and Crisp, 1946). Again, a high retreating angle is necessary for filling of the tracheae with gas in any insect with a closed tracheal system, according to the mechanism proposed by Wigglesworth (1953).

Changes in contact angle associated with the moulting cycle have been demonstrated by Wigglesworth (1947) and again by Holdgate (1955*b*) who has shown that in *Tenebrio* pupae the advancing angle rises from about 110° to 150°, which corresponds to a change from the fairly smooth early pupa to the rough surface of the mature pupa after the secretion of a surface wax. The latter is shown by the electron microscope to be in the form of closely packed, very fine filaments (Holdgate, Menter and Seal, 1955).

Because of the large changes in angle which can be caused by hydration and by surface roughness, irrespective of chemical composition, measurement of contact angles cannot be expected to provide any information about the nature of the surface materials. Low contact angles in aquatic insects, for instance,

30

do not imply the absence of a wax layer. Furthermore, even if the wax layer be removed, the underlying lipoprotein of the epicuticle may show a similar contact angle.

The significance of information about wetting of the cuticle is, firstly, that it throws light on certain ecological adaptations, and secondly, that it contributes to an understanding of the mechanisms of water-proofing and the formation of water-proof layers.

CHAPTER 3

EXCRETION AND OSMOREGULATION

THE main functions of excretory organs are the elimination of excess nitrogen and the maintenance of water and ionic balance in the body.

Excess nitrogen derives from two sources: protein and nucleic acid metabolism. Ammonia from protein catabolism may either be excreted as such or built up via the ornithine cycle or other mechanisms to urea and thence to uric acid. On the other hand, the products of nucleic acid catabolism, mainly purines, may also be excreted as such, or, more usually, degraded by various stages including uric acid, allantoin, allantoic acid and urea, to ammonia. The enzyme complement of many animals is such that the products of purine degradation and of synthesis from ammonia stop at the same compound after having reached it from different routes (Florkin and Duchateau, 1943).

It has been recognized for a long time that the end-products of nitrogen metabolism in animals are related to the availability of water. Thus many aquatic invertebrates are mainly ammonotelic, Amphibia and mammals are ureotelic, and birds, some reptiles and insects are uricotelic. This generalization has been supported by Delaunay (1931) and by Needham (1929, 1935) who lays emphasis on the environment of the egg stage in determining nitrogenous end-products.

In most terrestrial insects water is in short supply, and it is therefore not surprising to find that nitrogen is excreted as the non-toxic compound uric acid, which contains less hydrogen than any other nitrogenous excretory compound, and is only slightly soluble, so that water may be separated from it without much osmotic work and it may be excreted in crystalline form. Much more is known about excretion in insects than in other terrestrial arthropods, and it will be convenient to consider

32

them first. Earlier work on the physiology of arthropod excretion has been reviewed by Maloeuf (1938) and by Timon-David (1945). The review by Florkin and Duchateau mentioned above contains much information on the biochemistry of excretion in insects.

EXCRETION IN INSECTS. Many organs and tissues play a part in the complex of homoiostatic mechanisms collectively known as excretion, but in insects the Malpighian tubes are most important. Urine is formed in the Malpighian tubes, usually as a fluid secretion, the solid concretions of uric acid and urates separating out later as water is withdrawn in the rectum. In some insects, however, notably muscid pupae and adult mosquitoes, the tubes contain solid uric acid all the time, and it is not certain whether the cells alternate in function between secretion and absorption, or whether there are distinct cells for each process. In *Rhodnius* there is a clear division of labour: the upper two-thirds of each tube contain a clear fluid and the single layer of cells which compose the wall has a 'honeycomb' internal surface, whereas in the lower third, uratic crystals occur, and the cells are lined by a brush border (fig. 12). By ligaturing the lower (proximal) tube in two places some hours after the insect had fed, Wigglesworth (1931) found that uric acid crystallized out above the top ligature and the tube became distended, but between the ligatures the fluid remained clear and no distension occurred. This and other experiments suggest that fluid is secreted into the distal end of the tube and withdrawn from the proximal region, and this was in fact observed after injection of neutral red into the body cavity. Now Wigglesworth also found that the reaction of the upper tube contents was slightly alkaline (pH 7·2) while that of the lower tube was acid (pH 6·6), which led him to suggest that uric acid is secreted as the soluble acid salt of potassium or sodium into the upper tube and that base and water are reabsorbed in the lower segment, precipitating uric acid (fig. 13). The advantage of this cycle is, of course, that base and water, both of which may be in short supply, can be used repeatedly. Further absorption of water occurs in the ampullae at the proximal ends of the tubes (fig. 12) and also in the rectum.

Absorption of water in the hind-gut and rectum has been

observed in a number of insects (Wigglesworth, 1932). When the contents of the mid-gut pass into the hind-gut these are quite fluid, and they become progressively drier as they pass along. Desiccation is particularly marked in the rectum,

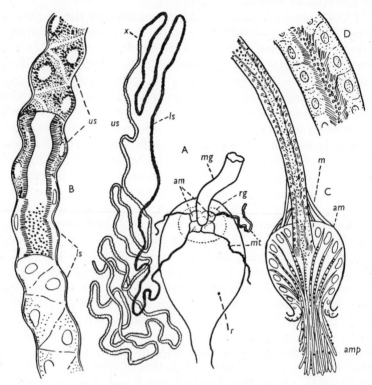

Fig. 12. A, excretory system of *Rhodnius*; only one Malpighian tube shown in full. B, junction of upper and lower segments of Malpighian tube, seen in surface view and optical section. C, lower end of Malpighian tube and terminal ampulla seen in optical section. D, detail of this region of tube. *am*, ampulla; *amp*, processes of ampulla cells; *ls*, lower (proximal) segment of Malpighian tube; *m*, muscle fibres; *mg*, mid gut; *mt*, Malpighian tubes; *r*, rectum; *rg*, rectal gland; *us*, upper (distal) segment of Malpighian tube; *x*, junction of upper and lower segments of tube. (From Wigglesworth, 1931.)

where the rectal glands are believed to play an important part in absorption. In *Lucilia*, indicator dyes accumulate round the bases of the rectal glands, pointing again to absorption of water (and possibly other substances) in that region. Patton and Craig (1939) demonstrated by means of radioactive sodium that reabsorption of salts taken up from the haemolymph into the Malpighian tubes of *Tenebrio* larvae occurs in the rectum. They believe that in this insect the entire reabsorption of water and utilizable material occurs in the rectum, the tubes only absorbing material from the haemolymph. In other insects, such as adult hymenoptera, the rectal contents are always fluid. If the rectal contents are not discharged for long periods, as in young honey bees for three weeks after emergence, and during the whole winter, water absorption in the rectum is clearly of importance.

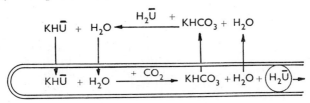

FIG. 13. Diagram showing possible mechanism by which free uric acid is precipitated in the lower segment of the Malpighian tubes in *Rhodnius*. (From Wigglesworth, 1931.)

In several insects, the distal part of the Malpighian tubes is intimately associated with the hind-gut in the region where water absorption is believed to occur. This has been observed in lepidopterous larvae (Henson, 1937) (fig. 14), Coleoptera (Lison, 1938; Patton and Craig, 1939) and in other orders. This cryptonephridial arrangement is probably a means for increasing the water-absorbing capacity of the rectal walls (Poll, 1938) rather than the removal of toxic substances as has been suggested. Patton and Craig (1939) demonstrated the absorptive powers of the distal portions of the cryptonephridial tubes in *Tenebrio* larvae by teasing a single loop free from the gut and enclosing it in a calibrated capillary tube filled with saline isotonic with the haemolymph. In these conditions there was no selective absorption of salt, water or amino acids, all of which entered the tube. Protein, however, was obstructed. Patay (1938) and Lison (1938) observed that at several points the wall of the perirectal tubes was very thin, and suggested that these areas, termed leptophragmata, facilitate water absorption by the tubes; but there is little reliable evidence as to their function.

The cryptonephridial arrangement is present in all higher lepidopterous larvae, save only in certain aquatic pyralids, which adds support to the belief that the system assists in water conservation.

In many Homoptera which feed on plant juices, water is in excess, and there are several more or less complicated arrangements whereby water from the fore-gut is short-circuited to the hind-gut. The hind-gut in these insects comes into intimate relation with the fore-gut, forming the so-called 'filter chamber'

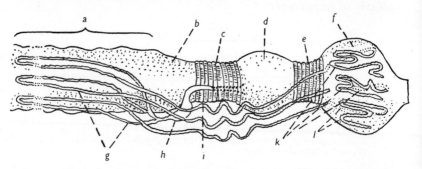

FIG. 14. Hind part of gut and Malpighian tubes of one side of the larva of *Vanessa urticae*: *a*, mid gut; *b*, ileum; *c*, anterior sphincter; *d*, colon; *e*, posterior sphincter; *f*, rectum; *g*, free part of Malpighian tubes; *h*, urinary bladder; *i*, free part of common duct; *k*, outer layer of tubes forming the rectal plexus; *l*, inner layer of tubes. (From Wigglesworth, 1950, after Henson, 1937.)

(Weber, 1930). In cercopids, the proximal portions of the Malpighian tubes lie coiled beneath the connective-tissue covering of a diverticulum of the oesophagus (Licent, 1912) where they presumably assist in extracting water.

EXCRETORY PRODUCTS IN OTHER ARTHROPODS. It is interesting to see whether the theory that the chief nitrogenous end-products of animals are determined by the availability of water applies within the arthropod phylum. In insects, we have seen that the product is uric acid. In the Arachnomorpha, the purine guanin, which is even less soluble than uric acid, is commonly found. Excretion in spiders was fully discussed by Millot (1926). Malpighian tubes are present, but according to Millot guanin is also excreted by cells in certain intestinal diverticula, and by the walls of the cloaca itself. Some of the guanin is deposited permanently in the integument. Peschen (1939) found that guanin forms more than 90 per cent of the nitrogenous excretory products in *Mygale*, the bird spider. But nothing is known of the physiology of formation of guanin in spiders.

In scorpions, Malpighian tubes, coxal glands and nephrocytes are the excretory organs, and guanin is the chief excretory product (Maloeuf, 1938). There are no Malpighian tubes in palpigrades, and the excretory product is unknown (Millot, 1942). Phalangids excrete guanin (Kastner, 1935). In the acarina, Malpighian tubes and coxal glands are present, and the end-product is guanin (Lees, 1946*b*). Arachnids therefore seem to conform to theory, at least they employ an insoluble end-product, and most of them are fully terrestrial. It would be interesting to know the end-product in the Xiphosura.

Turning now to the myriapods, we find that centipedes excrete uric acid by means of Malpighian tubes (Wang and Wu, 1948), and the distribution of uric-acid crystals is similar to that in the insect *Rhodnius*. Millipedes possess Malpighian tubes, but the excretory product is unknown (Folco, 1932).

Uric acid has been found as an excretory product in the Onycophora (Manton and Heatley, 1937). This is remarkable in view of the extremely hygric habitat and primitive nature of these animals. Uric acid is excreted by evacuation of a peritrophic membrane. Picken (1936) investigated the process of urine formation in *Peripatopsis* and concluded that filtration was hydrostatically possible, and that secretion and reabsorption almost certainly occurred in the nephridia. But he did not determine the main nitrogenous end-product.

The terrestrial isopods might be considered a test case for the theory relating habitat with nitrogenous end-product, for the various species are in general adapted to terrestrial conditions to different degrees (Edney, 1954). But Dresel and Moyle (1950) found that the main excretory product is ammonia, even in *Armadillidium*, the most terrestrial form studied, where it accounts for some 50–60 per cent of the non-protein nitrogen excreted. Some uric acid (less than 10 per cent) was found in the tissues, and rather more occurred in *Armadillidium* than in the less terrestrial *Oniscus* and *Ligia*, but, to complicate matters, still more uric acid was found in the related fresh-water *Asellus*.

37

Mödlinger (1934) found uric acid in three other strongly hygrophilic species, but made no quantitative measurements. The terrestrial isopods, then, do not seem to fit conveniently into the theory. Dresel and Moyle suggest that, since the total nitrogen excreted per unit weight is considerably less in terrestrial than in aquatic species, the main adaptation to life on land so far as excretion is concerned may have taken the form of a general reduction in nitrogen metabolism rather than by conversion from ammonotely to uricotely.

In general, however, it is clear that most terrestrial arthropods from dry habitats avoid loss of water by excreting nitrogenous waste in the form of uric acid or guanin. Most aquatic arthropods excrete ammonia (references in Prosser, 1952) and to this extent the phylum as a whole conforms to theory.

THE OSMOTIC PROBLEM. The kind of osmotic stress to which arthropods are subject depends, of course, on their environment and habits. Animals which feed on very dilute watery food, such as some plant-suckers, and those which live in fresh water, are concerned with preventing hydration of the blood and tissues; but for those which feed on dry food or live in salt water the problem is reversed.

Total osmotic pressure varies from one species to another and is usually rather high. Expressed as sodium chloride concentration, instead of the usual 0·9 per cent for mammals, values of over 1 per cent are common, rising to 2·12 per cent in *Tenebrio* larvae (Patton and Craig, 1939). A large part of this total osmotic pressure is contributed by amino acids, whose concentration approaches 20 per cent, or 50–100 times the normal concentration for mammals (Duval *et al.*, 1928). Salt concentrations are correspondingly reduced: chloride, expressed as sodium chloride, accounts for less than 15 per cent of the total molecular concentration in lepidopterous larvae and pupae, as compared with 65–70 per cent in mammals (Portier and Duval, 1927). Of the kations, Na^+ is often present in greater concentration than K^+, but the Na/K ratio is very variable, and is usually, but not always, much higher in carnivorous than in vegetarian insects. In *Cimex*, Na/K = 15·5, in *Melolontha*, 0·42 (Boné, 1945, 1947). Further information on these matters will be found in the standard texts.

When relatively sudden osmotic stress occurs, it is sometimes tolerated; but where the stress is less severe or abrupt, insects show remarkable powers of osmotic homoiostasis. In insects from dry habitats, and in salt-water forms, where the tendency is towards loss of water, this is counteracted by excretion of a hypertonic urine, by the use of endogenous water (p. 53) and, as we have seen, by the prevention of water loss. When excess water tends to enter the body, as in fresh-water insects and in plant-sucking bugs, the main regulatory process is excretion of a hypotonic urine. But these are generalizations, and there are many exceptions.

There is not a lot of information on the extent to which insects can tolerate rapidly rising osmotic pressures. A rise from 0·85 to 1·0 per cent sodium chloride equivalent occurs and is tolerated in *Culex* larvae after asphyxiation or struggling in the absence of oxygen (Wigglesworth, 1938). If the blood osmotic pressure is artificially increased by injecting hypertonic sodium chloride into the haemocoele of the cockroach, blood volume is rapidly increased (Munson and Yeager, 1949), presumably at the expense of tissue water. Later, these authors believe, some salt passes into the tissues, establishing a blood/tissue equilibrium at a higher level. If so, this is a case of tolerance rather than regulation. Similarly, *Anopheles* adults tolerate increases in the blood osmotic pressure up to 50 per cent above normal when this is caused by desiccation, according to Vinogradskaja (1936).

REGULATION IN TERRESTRIAL INSECTS. Osmotic and ionic unbalance may occur in an acute form in blood-sucking arthropods, when large quantities of vertebrate blood are ingested. This is usually rather hypotonic to their own haemolymph, but the salt concentration is much higher, and ionically very different (Lester and Lloyd, 1928; Wigglesworth, 1931; Mellanby, 1939a). In *Rhodnius* a large proportion of the water in a meal is eliminated by rapid excretion, and the same is true of *Glossina* where urine equal to 70 per cent of the total volume of the insect is produced in an hour (Lester and Lloyd, 1928). This diuresis serves to restore the insect's mobility rapidly and at the same time to eliminate excess salts. Removal of water from the gut is not selective, electrolytes are also extracted, so that the osmotic pressure of the gut contents is prevented from

39

rising so high as to withdraw water from the haemolymph. Total osmotic pressure and ionic content of the haemolymph are held reasonably constant by selective excretion, and this is reflected in the varying composition of the urine (fig. 15).

The power of ionic regulation of the haemolymph is illustrated by the experiments of Tobias (1948) on *Periplaneta*. This insect has a normal Na/K ratio of 6·2/1. By forced feeding with 1·4 M-KCl the gut was exposed to a solution of more than eight times the normal osmotic pressure and eighty

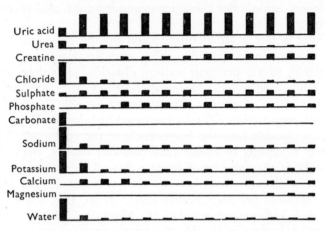

Fig. 15. The course of excretion of the chief urinary constituents during 13 days after feeding in *Rhodnius*. (From Wigglesworth, 1931.)

times the potassium concentration of normal haemolymph. In spite of this the animals survived, the sodium and water of the serum were unchanged and the serum potassium was only tripled. How much of this regulation was achieved by the gut lining and how much by the excretory system is unknown.

REGULATION IN AQUATIC INSECTS. In fresh-water insects there is a danger of losing salts to the environment, and of gaining water from it. For insects living in sea water the problem is reversed, while in brackish water, although osmotic stress may be reduced, it is often variable, and ionic ratios are usually different from those of the body fluid.

In general, aquatic insect larvae appear to be homoiosmotic over a limited range of environmental conditions, but the means of achieving this are varied. Adult aquatic insects which breathe atmospheric air may maintain their osmotic equilibria in the same way as terrestrial animals if their integuments are impermeable to water and salts. Krogh (1939) believed this to be true for diving beetles. There is some evidence, however, that *Dytiscus* and others are not homoiosmotic in hypertonic media (Bachman, 1912). Claus (1937) found an active regulatory process in three species of *Sigara* (*Corixa*). *S. lugubris*, a brackish-water species, was homoiosmotic over its natural range of 0·1–1·5 per cent salinity, and hypotonic to the medium on higher salt concentrations, but the fresh-water species were strictly homoiosmotic only in pond water.

In truly aquatic (non-air-breathing) insects it might be thought that the respiratory surfaces would lead to osmotic problems. This is not the case, apparently, in *Sialis* larvae (Beadle and Shaw, 1950). In this insect the concentration of plasma chloride expressed as sodium chloride is only 0·15–0·35 per cent out of a total osmotic pressure equivalent of 0·8–1·0 per cent sodium chloride. After as long as five weeks in distilled water, the chloride fraction is only reduced to 0·1 per cent, and this can be restored in a few days if the larvae are transferred to 1·0 per cent NaCl. The cuticle is therefore nearly impermeable to the outward passage of Cl⁻, and this is the animal's main line of defence against salt loss. Shaw (1955*b*) has recently confirmed the low permeability of *Sialis* cuticle to both Cl⁻ and Na⁺. Beadle and Shaw also found that if the larvae were bled so that up to half the total chloride was lost and then replaced in tap water, much water was absorbed to restore the plasma volume, but no active absorption of chloride occurred. Nevertheless, the total osmotic pressure of the plasma was kept constant by changing the non-protein nitrogen content to compensate for the loss of chloride. This was apparently achieved at the expense of plasma protein which may be reduced to 1/20th of the normal level without affecting the larva—a remarkable example of tolerance. Shaw (1955*b*) has recently taken this story a good deal further in an investigation of the regulation of Na⁺, K⁺ and Cl⁻ in *Sialis*. This work is discussed below (p. 47).

Wigglesworth (1938) also found that in *Aëdes aegypti* and in *Culex pipiens* depletion in chloride content of the haemolymph

41

was compensated for by changes in non-chloride solutes so that the total osmotic pressure remained more or less constant. Treherne (1954a) found a similar state of affairs in larvae of the beetle *Helodes*.

In other aquatic larvae, however, active uptake of chlorides does appear to take place. In mosquito larvae the anal papillae, which were at one time thought to be respiratory organs, are concerned in the absorption of water and salts. The anal papillae are the only areas of the body freely permeable to both salts and water, for *Aëdes* larvae, which shrink in hypertonic solutions, do not do so if the anal papillae are ligated or destroyed (Wigglesworth (1933b, c). Koch (1938) showed by chemical analysis that in both *Chironomus* and *Culex* larvae salt is lost to salt-free water, but that both animals can absorb chlorides from solutions as dilute as 0·001 M—unless the anal papillae are destroyed, when a slight loss is sustained. Treherne (1954b) used labelled Na^+ to demonstrate that uptake of this ion occurs largely through the anal papillae in *Aëdes aegypti* larvae. He also found that sodium uptake was independent of potassium concentration, and suggested that separate mechanisms control the uptake of these ions.

There is a good general correlation between environmental salinity and development of anal papillae in different species of mosquito larvae (Gibbins, 1932); and Wigglesworth (1938) showed that in larvae experimentally adapted to increasing salinities the papillae become progressively reduced in size (fig. 16). The ability to take up chlorides is affected by the previous history of the larvae, for Wigglesworth (1938) showed that larvae reared in low chloride concentrations retain salt better in distilled water and take it up faster from dilute solutions than those reared in higher concentrations.

Active absorption of this kind has been demonstrated in several other insects: in *Libellula* and *Aeschna* larvae by Krogh (1939), where absorption seems to occur through the rectal gill plaques, and in *Corethra* larvae by Schaller (1949). Hers (1942) confirmed the active nature of such uptake when she found that *Chironomus* larvae, deprived of salt by being kept in distilled water, were only able to replace it from tap water if this were oxygenated.

It is commonly found that insects from salt water are capable

42

of much better osmoregulation than related species from fresh water. Beadle (1939) found that in the salt-water *Anopheles detritus*, although both chloride concentration and total osmotic pressure of the haemolymph increase in strongly hypertonic media they remain well below that of the environment (fig. 17). In the fresh-water *Aëdes aegypti*, however, both values rise as much as the external medium when the latter is above about

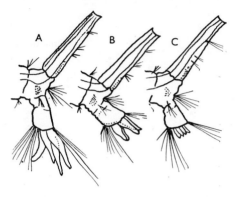

FIG. 16. Posterior extremity of larvae of *Culex pipiens* showing variation in size of anal papillae. A, reared in distilled water; B, in tap water with a chloride content equivalent to 0·006% NaCl; C, in salt water equivalent to 0·65% NaCl. (From Wigglesworth, 1950.)

1·0 per cent salinity (Wigglesworth, 1938, and fig. 17). In low concentrations both these species can maintain their haemolymph hypertonic to the environment (by means of non-chloride solutes). Ligation experiments showed that in *A. detritus* salt and water uptake occurred exclusively across the gut wall. This species, therefore, has an impermeable integument and its osmotic problems arise solely from imbibing water with its food.

THE SITE OF OSMOTIC WORK. Although the above work clearly demonstrates active uptake, it provides no evidence as to where the osmotic work is done, and this aspect of the problem has recently received attention. By means of a refined method of freezing-point determination (Ramsay, 1949) it has been possible to measure the total osmotic pressure of small samples of fluid, and thus accurately to compare the tubal fluid with the haemolymph and the rectal fluid in this respect. Ramsay (1950) found in this way that for *Aëdes aegypti* in

43

distilled water, fluid in that part of the intestine next to the Malpighian tubes (which is very probably tubal fluid itself) is isotonic with or feebly hypertonic to the haemolymph, while fluid from the rectum is strongly hypotonic to it. This suggests absorption of solutes by the wall of the rectum. In *Anopheles*

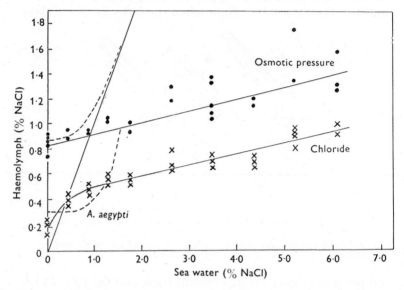

Fig. 17. Changes in osmotic pressure and chloride concentration in the haemolymph of *Anopheles detritus* (full lines) and *Aëdes aegypti* (broken lines) in various strengths of sea water. The straight line through the origin represents concentrations where the medium is isotonic with the haemolymph. (From Beadle, 1939.)

detritus in sea water the opposite occurs, and rectal fluid becomes hypertonic to the haemolymph and approximately isotonic with the external medium. If *A. detritus* is kept in fresh water, however, the rectal fluid is hypotonic to the haemolymph, so that in this species absorption of either salt or water by the wall of the rectum, as required, occurs.

In *A. aegypti*, Ramsay also found an internal cycle of water round the circuit, haemolymph→mid-gut→intestinal caeca→haemolymph; but that, of course, does not affect exchange with the environment.

Now, although the tubal fluid has been shown to be approximately isotonic with the haemolymph, this is not a demonstration that the tubes play no part in *ionic* regulation. In fact it appears that they do (Ramsay, 1951) for the tubal fluid of *A. aegypti* in fresh water contains only half the amount of sodium present in the haemolymph. In larvae kept in sodium chloride solution, the concentration of sodium in the tubal fluid increased, until in 1·0 per cent sodium chloride it was equivalent to that of the haemolymph, but never rose above it. Thus, the tubes do play a part in salt retention in distilled water, but they do not actively remove sodium from the haemolymph in hypertonic media. No evidence of a change in sodium concentration along the length of the tubes was found.

According to Boné and Koch (1942), the Malpighian tubes of *Limnophilus flavicornis* are capable of producing a urine hypertonic to the blood. In this respect *Limnophilus* differs from *Aëdes*. By withdrawing samples of blood for analysis from *Limnophilus* living in different hypertonic media Boné and Koch were able to show that the intensity of chloride elimination by the tubes kept pace with the rate of uptake from the environment, so that the total blood chloride remained constant.

IONIC REGULATION. An improved technique of flame photometry allowing simultaneous determination of sodium and potassium in very small quantities of fluid was developed by Ramsay, Brown and Falloon (1953). Using this method Ramsay (1953*a*) took the study of salt and water balance in mosquito larvae a step farther. Both sodium and potassium, he found, enter the body together with water by the anal papillae, and when either ion is present in the medium in excess, its concentration in the tubal fluid increases and reabsorption from the rectum decreases in such a way that the composition of the haemolymph remains relatively constant in face of changes in the medium (Table 2).

As Ramsay points out, it is remarkable that whereas the sodium concentration of the intestinal fluid never exceeded that of the haemolymph, potassium concentration always did, irrespective of the composition of the medium. Potassium is reabsorbed in the rectum or mid-gut, thus a continuous circulation of potassium must take place (p. 47).

A somewhat similar state of affairs obtains in *Rhodnius* (Ramsay, 1952). In this insect the Na/K ratio in the haemolymph lies normally between 10/1 and 20/1. After a normal feed, the urine at first contains more sodium than potassium, but after some hours the potassium concentration exceeds that of sodium. The upper portion of the Malpighian tubes always contains less sodium and more potassium than the haemolymph and the total osmotic pressure is greater than that of the haemolymph. In the lower part of the tubes these differences are reduced, certainly by the absorption of potassium from the tube into the haemolymph and possibly by the absorption of sodium also. Water is probably absorbed at the same time. In the rectum, water and sodium are probably absorbed by the ampullae or rectal glands, and the osmotic pressure of the rectal contents rises considerably.

TABLE 2

(Data from Ramsay, 1953*a*.)

(*a*) *Mean concentrations of sodium and potassium (in m.equiv./l.) in the haemolymph of* Aëdes aegypti *in different external media*

Distilled water		NaCl		KCl	
		85 mM/l.	1·7 mM/l.	85 mM/l.	1·7 mM/l.
Na	K	Na	Na	K	K
87	3·1	113	100	5·7	4·2

(*b*) *Mean concentrations of the intestinal fluid (believed to be tubal fluid) (in m.equiv./l.) in different external media*

Distilled water		NaCl, 85 mM/l.		KCl, 85 mM/l.	
Na	K	Na	K	Na	K
24	88	71	90	23	138

Throughout the fluctuation in sodium and potassium content of the urine which occurs during the period of diuresis, the composition of the haemolymph remains relatively constant. The tubes and rectum together are therefore capable of regulating both the total osmotic pressure and the ionic composition of the haemolymph.

In *Sialis* larvae the maintenance of normal Na^+, K^+ and Cl^- concentrations in the blood has also been found to depend upon regulation by the tubes and rectum in producing urine of varying composition (Shaw, 1955*b*). In normal larvae none

46

of these ions is present in the urine; excess of K^+ or Na^+ taken in through the gut is excreted in the urine, rapidly in the case of potassium. (Potassium has a detrimental effect upon the neuro-muscular system of the larva if it is in excess.) Cl^-, however, tends to be conserved, perhaps because there is no means of active absorption of this ion. Water is absorbed osmotically through the cuticle. It may also be absorbed through the gut wall together with sodium for which there is an active process of uptake. Normally water output by the excretory system balances uptake, but the presence of extra sodium inhibits water loss, and Shaw suggests that sodium may act as a regulator in this respect.

In hypertonic media these larvae cannot survive, because osmotic water loss and uptake of salts through the gut could only be offset by the production of a hypertonic urine, and although salts can be concentrated in the rectal fluid, hypertonicity is never reached.

POTASSIUM CIRCULATION. In *Rhodnius* and in *Aëdes* larvae (Ramsay, 1952, 1953*a*), in *Sialis* larvae (Shaw, 1955*b*), and in a number of other insects of different orders, different feeding habits and from both aquatic and terrestrial environments (Ramsay, 1953*b*), a circulation of potassium has been found, from haemolymph to tubule to rectum or intestine and back to haemolymph, irrespective of potassium concentration in the haemolymph. Furthermore, in most of these insects, the passage of potassium across the tubule wall from the haemolymph has been shown to be against an electro-thermal gradient, and in that sense truly active (Ramsay, 1953*b*). This led to the suggestion that potassium circulation might play a necessary part in urine formation: perhaps its active transport into the tubule led to an osmotic gradient along which sodium and water could follow passively.

But water itself has now been shown to be actively transported into the tubule, at least in *Dixippus* (Ramsay, 1954), so that this cannot be the reason for potassium circulation.

Wigglesworth's original suggestion (1931) that circulation of a base (he chose potassium because its urate is more soluble than that of sodium) is necessary for uric acid transport across the tubule wall may still be true, though, as Ramsay points out, the much greater amount of potassium circulated than

47

that of uric acid excreted would seem to tell against this inter-pretation.

Perhaps potassium is necessary in small amounts for this purpose, and the mechanism having been developed, its extent was exaggerated as a handy means of increasing osmotic pressure in the tubule and thus reducing the osmotic work involved in getting water into the tubule. But there is much to be learnt in this field before a clear picture emerges.

OSMOREGULATION IN OTHER ARTHROPODS. Chloride and water regulation in the argassid tick *Ornithodorus moubata* was investigated by Boné (1943). In this animal a large quantity of mammalian blood is ingested in a short time, and the excess water is eliminated by the coxal glands. Boné showed that mammalian blood, whose chloride content is approximately equivalent to 0·5 per cent NaCl, is concentrated in the gut by the absorption of water and salts into the haemocoele. The haemolymph has a chloride concentration of about 1·0 per cent NaCl and this is maintained by the excretion of excess salts through the coxal gland system. The coxal fluid is never hypertonic to the haemolymph: it has a normal chloride concentration of about 0·8 per cent NaCl, but it varies according to the haemolymph concentration. The coxal glands therefore function as both osmotic and ionic regulatory organs.

Lees (1946*b*) confirmed Boné's figures, and advanced reasons for believing that the mechanism of coxal fluid formation is by filtration and reabsorption of salts. This is supported by the fact that the coxal glands are distensible by a muscle, leading to reduced internal pressure; their walls are only 1–2 μ thick, whereas the tubal wall, through which reabsorption probably occurs, is 5·3 μ thick. Furthermore, there is a very rapid passage of injected dyes from the haemolymph into the coxal fluid. Serum albumin sometimes passes after injection, but normal haemolymph proteins never do.

Ixodes ricinus behaves in a very different manner (Lees, 1946*b*). There are no coxal glands and feeding is very pro-longed, so that time is available for the elimination of water through the integument, and perhaps to a small extent through the Malpighian tubes. There are no special regulatory pro-cesses, and the tick tolerates wide variations in haemolymph concentration. After desiccation, for example, this rises to 1·3 per cent NaCl.

In both these animals, the products of nitrogen metabolism (mostly guanin) are excreted by the Malpighian tubes. Unlike insects, therefore, the functions of nitrogen excretion and osmoregulation are carried out by separate organs. It would be very interesting to know whether this is also true in other arachnomorph animals.

Information on osmotic and ionic regulation in terrestrial isopods has been reviewed by Edney (1954). In general, it appears from the work of G. Parry (1953) on *Ligia oceanica* and other species that osmotic tolerance rather than regulation is characteristic of this group of animals. *Ligia* can regulate its haemolymph osmotic pressure indefinitely in 50–100 per cent sea water; in lower or higher concentrations, however, haemolymph osmotic pressure follows that of the medium. The normal osmotic pressure for *Ligia* lies between Δ 1·98 and 2·23° C., but depressions as low as 1·44° C. and as high as 3·75° C. can be tolerated. There are significant changes in total osmotic pressure during the moult (Widmann, 1935; G. Parry, 1953), and these are probably associated with calcium mobility at that period. Bursell (1955) showed clearly that transpiration in *Oniscus* leads to an increase in osmotic pressure of the haemolymph, and the animals are not capable of regulating blood volume in these circumstances.

CONCLUSIONS. In the vast majority of terrestrial arthropods water is in short supply for the greater part of the time, and prevention of water loss as a result of excretion is necessary. This is accomplished by excreting nitrogen in the form of relatively insoluble compounds: uric acid (insects and centipedes) or guanin (arachnomorphs), and water is retained by an impermeable cuticle and rectal reabsorption.

For salt-water larvae the problem is essentially similar, and is met by the excretion of a hypertonic urine (*Anopheles detritus*). No case of extraction of water alone from a saline environment has been reported (save perhaps in some insects eggs (p. 71)).

Where excess water is the problem, it may be met by impermeability of the cuticle, as in adult fresh-water insects and *Sialis* larvae, or by the excretion of a hypotonic urine and retention of electrolytes as in most fresh-water larvae. Salt uptake is accomplished by feeding (*Sialis* and terrestrial insects) or by uptake together with water through permeable regions of the integument such as the anal papillae of mosquito larvae, *Chironomus*, etc. and the rectal gills of *Libellula*. Again, there has been no demonstration of selective absorption of salts from aquatic environments; water and salts are taken up together, the latter retained by the Malpighian-tube and rectal system, and the former excreted.

Most aquatic larvae are homoiosmotic in hypotonic media, but their ability to remain so in hypertonic situations varies, those from salt water being more efficient in this respect. Terrestrial species, too, vary in the degree of homoiostasis which they attain. *Rhodnius* and *Ornithodorus* are homoiosmotic after the osmotic stress of a large blood meal; but if the stress is unusual, considerable changes in salt concentration may be tolerated (*Periplaneta*, adult *Anopheles*).

As regards the internal mechanism of regulation, there are two main systems. First the amino acid/serum protein balance, which is capable of counteracting large variations in total osmotic pressure caused by variation in the salt content. Secondly, the Malpighian-tube and rectal system, which regulates the total osmotic pressure and ionic balance. Thus tubal urine may differ in solute concentration from the blood according to salt and water requirements of the moment. In *Tenebrio*, Patton and Craig (1939) found no solute selection, but such has been demonstrated in *Limnophilus* and *Aëdes*. In the rectum, water or salts or both may be reabsorbed from the urine into the haemolymph, again according to need. Thus, apart from the absence of filtration, the arthropod Malpighian-tube and rectal system is analogous to the vertebrate glomerular kidney.

The internal circulation of large amounts of potassium discovered by Ramsay may be associated with uric acid and/or water transport into the tube, but its significance is by no means clear.

Taken as a whole, the terrestrial arthropods show regulatory mechanisms analogous to those of mammals, if within wider limits. They differ from mammals in the greater degree of tolerance of accidental or forced changes which they show.

CHAPTER 4

GAIN OF WATER

EXCEPT in a very few cases, arthropods living in air unsaturated with water vapour transpire continuously, albeit sometimes very slowly. In order to maintain a balance water must be gained, and it is the purpose of this chapter to consider the various ways in which this may occur.

INTAKE WITH THE FOOD. The food of some insects such as aphids consists very largely of water, and they excrete a very copious watery urine. When feeding is intermittent, desiccation during a non-feeding interval may lead to the retention of a greater proportion of water in the next meal—a form of regulation found by Mellanby (1932b) in *Cimex*. Insects such as the honey-bee and muscid flies, which produce a liquid urine, drink frequently to replace the water lost. But for insects which live in really dry surroundings without access to liquid water, the water content of the food is important and may determine the limits of their distribution. *Ephestia kuehniella* and *Tenebrio molitor* can live on food containing only 1 per cent water, *Tribolium*, *Silvanus* and others need at least 6 per cent, and *Lasioderma*, *Sitodrepa* and *Ptinus* require 10 per cent (references in Wigglesworth, 1950). Arthropods living in the desert feed upon fragments of dead plant material which have absorbed water from the atmosphere during the night (Buxton, 1924a). *Tenebrio* larvae probably consume more food than they need, for the sake of its water content (Schultz, 1930).

ORAL AND ANAL DRINKING. The drinking of liquid water has been reported in a large number of insects (Leclercq, 1946), though the habit is confined to adults; larvae obtaining their water with the food. It may well be that many arthropods drink water from a moist surface against a suction pressure. This has not been demonstrated in any insect, but it certainly occurs in spiders and woodlice. D. Parry (1954) examined the capacity of two species of lycosid spiders to imbibe capillary water from soils. Using carborundum powders of different particle size, and applying suction pressures to the capillary water in them, he showed that these spiders are able to drink against pressures up to 600 mm. mercury. The rate of drinking fell off with increased suction and finally ceased, even though the soil might remain saturated. It did not fall off with time, and was unaffected by soil particle size at a given suction.

Millot and Fontaine (1937) believe that spiders fall into two groups: first, a large group the members of which drink water, have a water content greater than 70 per cent of the total body weight and a high (but unproved) cutaneous permeability to water; secondly, a small group with the opposite properties. The normal water content, they report, is very constant for any one species.

Both oral and anal uptake of water have been observed in woodlice (Spencer and Edney, 1954). In the littoral species, *Ligia oceanica*, most of the uptake was anal, in other more terrestrial species it was by mouth. All except *Ligia* could imbibe water by mouth from moist plaster of Paris sufficiently rapidly to offset transpiration from the integument in moving air at 85 per cent R.H. This faculty is probably employed in nature, for in all but saturated air woodlice transpire rather rapidly. No absorption through the integument was demonstrated.

ABSORPTION THROUGH THE CUTICLE. The post-embryonic stages of several insects possess special areas for the absorption of water. Aquatic larvae form a special case, considered elsewhere (p. 42): here we are interested in terrestrial forms. Nutman (1941) reported that in the collembolan *Onychiurus* the ventral tube is the main water-absorbing organ, and Schneider (1948) found that desiccated larvae of the syrphid *Epistrophe* extend the anal papillae rhythmically into a drop of water and absorb up to 50 per cent of their weight in a few hours. Probably the whole of the cuticle of most arthropods is capable of taking up water to some extent. There is no osmotic barrier and even a waxed cuticle is slightly permeable. Such cutaneous absorption has been demonstrated in *Phlebotomus* larvae (Theodor, 1936) and in *Acridium* (Colosi, 1933).

The cuticle of *Agriotes* larvae from the soil is abraded, and

52

the insects then become more freely permeable to water. Evans (1944) expressed the suction power of soils on the pF scale, which enables the effects of different suction pressures to be compared irrespective of particle size, and found that *Agriotes* larvae came into equilibrium with soil of pF 3·9, which is equivalent to 0·33 M sucrose.

METABOLIC WATER. It has been suspected for many years that insects living permanently in dry surroundings manage to survive by retaining the water derived from complete oxidation of their food (Berger, 1907; Babcock, 1912). In 1930 Buxton showed that *Tenebrio* larvae, starved at different relative humidities below 60 per cent, lost more weight in lower humidities than in high, but that the proportion of dry to wet weight remained constant. This he explained by supposing that, in dry air, additional dry material was oxidized and the water of metabolism retained to compensate for that lost by evaporation. (The complete oxidation of 100 g. of carbohydrate yields about 60 g. water; 100 g. of fat yields 107 g. water.)

Mellanby (1932a) analysed the food and water reserves of *Tenebrio* larvae starved at different temperatures and humidities. He found that the proportion of fats to other foods used was such that the water produced almost exactly equalled the weight of carbon lost, and since the animals did not excrete, any loss in weight represented water lost by evaporation. At 23° C. he found that metabolism was regulated so that at all humidities the wet/dry weight ratio remained constant. At 30° C., however, metabolism was higher, and at the same rate in all humidities. Because of this, water was lost at high humidities less rapidly than it was gained by metabolism, and the larvae in consequence gained weight. At 37° C. the effect was even more pronounced, and occurred down to 70 per cent R.H. (Absorption of water vapour above 90 per cent R.H. is another matter and will be mentioned below.)

Buxton and Lewis (1934) found that in starving *Glossina*, additional fat was oxidized in dry air, 'presumably in order to produce water of metabolism to compensate for the excessive evaporation'. Yeager and Munson (1950) found that after 35 days' starvation *Periplaneta* nymphs suffered no change in the proportion of blood volume to total weight, though the latter had fallen by 38 per cent. A similar regulation of water content and total osmotic pressure is reported by Ludwig and Wugmeister (1953) in starved *Popillia*.

The work so far mentioned is at least consistent with the idea that water obtained by the metabolism of dry food material is sufficient to counteract desiccation, and that in some cases, drier conditions apparently stimulate greater metabolism. On the other hand, contrary results have been reported. Mellanby (1932*b*, 1934*a*) failed to observe increases in the rate of metabolism in fasting *Cimex* or *Tineola* larvae at low humidities. Buxton (1932*b*) found that *Rhodnius* used *less* solid matter in dry than in moist air, and Gunn and Cosway (1942) found no effect of humidity on the respiration rate of cockroaches, and concluded that no extra water was produced as a result of the desiccated condition.

Fraenkel and Blewett (1944) approached the problem by measuring the ratio between dry weight of food eaten by the larvae and dry weight of the resulting pupae of *Tribolium*, *Ephestia* and *Dermestes* at different humidities. Their figures show clearly that in low humidities more dry food is required to produce a given weight of pupa than in high humidities. In drier air the larval period was longer than in moister air, and the resulting pupae were smaller, but the proportion of water in pupae from larvae in all humidities was the same. Speicher (1931) had earlier found the same phenomenon in *Ephestia*.

Now it is clear that the experiments quoted have not all been designed to answer the same question. If the question is simply 'does the animal retain and use the water derived from metabolism?', the answer must be that it does, in so far as the animal concerned can be said to retain water at all. To prove this it is only necessary to show that the sum, 'water lost + water present', is greater than the amount of water taken in with the food (assuming that this is the only source of water intake). This has in fact been demonstrated by Fraenkel and Blewett (1944).

A further question may be asked: 'do insects metabolize more dry material in dry air than in moist?' This is an interesting question, and there is evidence that some do, namely *Ephestia*, *Tribolium*, *Dermestes* (Fraenkel and Blewett), *Tenebrio* (Buxton, Mellanby); while others do not, *Cimex*, *Tineola* (Mellanby), *Rhodnius* (Buxton), *Periplaneta* (Gunn and Cosway). Where such increased metabolism occurs, presumably the extra water resulting is available for other purposes, e.g. to supplement that lost by evaporation. Where it does not, it cannot be concluded that metabolic water is not available for use by the animal, but only that *extra* metabolic water is not produced as a result of desiccation.

A different question again is whether an animal can maintain a constant wet/dry weight ratio in the face of a drop in total weight when starved in dry air. If the ratio falls, that is evidence that water lost is greater than water gained, but not that no water is gained from metabolism. If the ratio is maintained, that is evidence that more dry material has been metabolized than would have been the case in moist air, and this must have provided some water. The balance, however, may have been struck by an increase or decrease in the rate of transpiration, or in the amount of water in the urine, not necessarily by the amount of dry material used. The significance of maintaining a constant dry/wet weight ratio in different humidities is that a balance is struck between water gained and water lost. Water of metabolism necessarily enters into this balance, but we cannot say whether it is the controlling factor when loss by transpiration, loss by excretion and loss through the spiracles may all vary directly or indirectly as a result of humidity conditions.

ABSORPTION OF WATER VAPOUR. There are a few well-authenticated cases of arthropods gaining water by absorbing water vapour from the air, but the mechanism of this process is as yet imperfectly understood.

It is important to scrutinize the evidence rather closely, for the difficulties involved in setting up and maintaining humidities near to saturation are great, and unless strict precautions are taken the humidity to which an animal is in fact subjected may be different from that intended by the experimenter. It would seem best to discount demonstrations of absorption of water vapour from saturated air, for when the air is fully saturated, local condensations of liquid water are unavoidable, and absorption may occur by the normal process of drinking the liquid, or by absorption of liquid water through the cuticle. We may first review the occurrence of this phenomenon in arthropods and then discuss possible mechanisms.

The first well-documented case is that of *Tenebrio* larvae, reported by Buxton in 1930, and confirmed by Mellanby (1932*a*). These larvae when starved were able to regulate their water content in air below about 80 per cent R.H., but in 88 per cent or above they gained weight and their water content rose. It has already been noticed that these insects gain water as a result of metabolizing dry food, but the amount of water gained in high humidities was significantly more than could be accounted for even by the metabolism of fat, and since they were starved, absorption of water vapour from the air is the only possible conclusion. Mellanby (1932*a*) reported that equilibrium was established at 88 per cent R.H. at several temperatures; below this the larvae lost weight. The process is therefore dependent upon relative humidity, not saturation deficit. Gain of weight by starved *Tenebrio* larvae has also been reported by Lafon and Teissier (1939) who found that the process continued for 12–15 days in saturated air, after which the weight remained constant for a long time. It has also been confirmed in unsaturated air by the writer (unpublished) and by T. Browning (personal communication), both of whom found the process to be irregular and intermittent.

Breitenbrecher (1918) claimed to have demonstrated uptake from moist air by *Leptinotarsa*, and Bodine (1921) for hibernating *Chortophaga* nymphs, but both these claims must be suspect, for the insects were either in contact with moist soil or in saturated air. The same may be said of Hodson's (1937) results with *Leptocoris*.

Kalmus (1936) found that the air in a small chamber containing *Tenebrio* larvae reached a relative humidity of about 90 per cent and remained in equilibrium at this level until the animals died, when it became saturated.

Mellanby's work on *Tineola* (1934*a*) showed an increase in the proportion of water present under certain conditions, but there is no need to suppose uptake from the air to explain this, for the absolute water content fell.

Another well-documented case is that reported by Ludwig (1937) who found that nymphs of the grasshopper *Chortophaga* starved at relative humidities above 95 per cent gained weight (fig. 18). All of his insects died of starvation after six days regardless of humidity, and at death the proportion of water had risen in those from humidities above 82 per cent.

There is no doubt that the faculty is present in ticks. Lees (1946a, 1947) clearly demonstrated the ability of unfed *Ixodes ricinus* and other species to extract water vapour from air at relative humidities down to 92 per cent (fig. 19). The uptake of water in *Ixodes* does not cease if the spiracles are blocked (respiration then appears to be cutaneous), but it does if the ticks are asphyxiated, cyanided or unduly desiccated. It appears

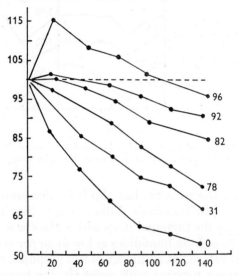

FIG. 18. Rate of loss of weight in *Chortophaga viridifasciata* at the different relative humidities indicated at the end of each curve. Ordinate: weight as percentage of original weight; abscissa: time in hours. (From Wigglesworth, 1950, after Ludwig, 1937.)

therefore that the process occurs over the whole surface of the animal and is not a purely physical one. Lees suggested that the epidermal cells play an important part in water uptake. T. Browning (1954b), working with *Ornithodorus moubata*, demonstrated the same faculty, and showed that carbon dioxide anaesthesis, which occurs in 35 per cent CO_2 in air, inhibited uptake. Blocking the spiracles of this species also prevented uptake. Furthermore, the rate of transpiration in dry air was greatly increased by anaesthesia in carbon dioxide, but not at the concentration necessary to open the spiracles (5 per cent).

Since the ability of ticks to absorb water from the air and also their ability to restrict water loss have thus been shown to be inactivated by the same carbon dioxide concentration, which is that necessary to produce anaesthesia, Browning's results add weight to Lees's suggestion that both processes are physiological ones, and perhaps both dependent upon epidermal activity.

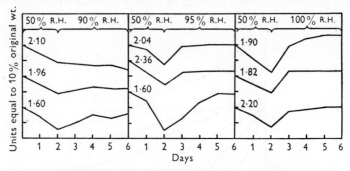

FIG. 19. Uptake of water at high humidities by unfed females after desiccation for 2 days at 50% R.H. (From Lees, 1946a.)

Absorption from the air has also been demonstrated in the rat flea *Xenopsylla brasiliensis* (Edney, 1947) (fig. 20). In this insect it is only the prepupal stage which absorbs water vapour, and it may do so from humidities as low as 50 per cent, irrespective of temperature. If a number of flea prepupae are enclosed in a small space together with moist air, the humidity falls to 50 per cent, remains there until the fleas pupate, and then rises again as the pupae lose water.

There is, then, good evidence that at least some arthropods are capable of extracting water vapour from unsaturated air, but the faculty is certainly not present universally, at least in such a marked form as the examples quoted above. It does not occur in several insects where it has been looked for, neither does it in woodlice (Spencer and Edney, 1954). The possibility must not be overlooked, however, that the mechanism, whatever it is, may be more widespread in a less well-developed form. In ticks it appears that secretory activity of the epidermal cells is involved in absorption, and it has been suggested by Lees that their activity also helps to reduce water loss. Certainly the fact that dead ticks lose water more rapidly than living ones is significant in this respect. In spiders, too (Davies and Edney, 1952), transpiration is greater from dead than from living animals. If we assume that living epidermal cells exercise a secretory activity which, if large enough, results in the absorption of water, but which if less marked, at least has the effect of restraining water loss, these facts would be explained.

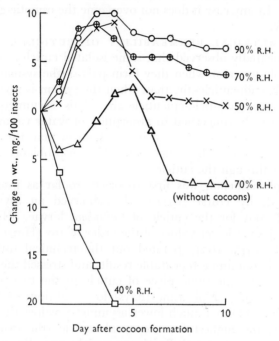

FIG. 20. Change in weight caused by absorption (or loss) of water vapour by *Xenopsylla* prepupae. The prepupal period lasts for 3–4 days after spinning. (After Edney, 1947.)

It is not unlikely that cells capable of secreting inwards should restrain transpiration outwards, but, as Beament (1954) has pointed out, this is no proof that they do.

The difficulty in providing a satisfactory explanation of the mechanism of this process lies in the fact that water is being moved against a very considerable osmotic gradient. A solution with an osmotic pressure equivalent to that of the haemolymph of a typical arthropod (about 1·0 per cent sodium

59

chloride) is in equilibrium with air at about 99 per cent R.H. Water will only be taken up from air at 90 per cent R.H. by a solution whose osmotic pressure is 140 atmospheres or more, which is equivalent to a 17 per cent sodium chloride solution. Mellanby (1932*a*) thought that water might be taken up by condensation in the tracheoles, but this is certainly not the explanation in ticks, where uptake occurs through the general surface. In any case it does not overcome the osmotic gradient difficulty.

ASYMMETRICAL PERMEABILITY OF CUTICLES. It has been repeatedly observed that some isolated cuticles are more permeable to water when they form part of the system water-epicuticle-endocuticle-air, than when the system is reversed to give water-endocuticle-epicuticle-air. This asymmetry may not be directly concerned in absorption of water vapour from air, but an understanding of its mechanism may throw light on the structure of the cuticle, and thus on problems of water transport through the cuticle.

The phenomenon was first reported, so far as insects are concerned, by Hurst (1941), who observed an asymmetry value of 100*x* for the cuticle of *Calliphora* larvae. He subsequently quoted lower values of the order of 10*x* (Hurst, 1948). Beament (1945, 1954) pointed out the technical difficulties involved in obtaining repeatable results and stressed the importance of using the same piece of cuticle in the two positions and of allowing ample time (days) for conditions to become stable. He obtained much lower asymmetry values than those of Hurst; the highest for a natural cuticle (the exuviae of *Rhodnius*) was 5·5*x*, and *Periplaneta* cuticle gave lower values still. Artificial membranes made by treating a de-waxed cicada wing with *Rhodnius* wax, or a tanned gelatin membrane with beeswax, gave asymmetry values of 4*x*. Boiling in chloroform raised the transpiration rate greatly and destroyed asymmetry.

Richards, Clausen and Smith (1953) investigated the same phenomena in the larvae of *Sarcophaga bullata*. By means of heavy water they were able to show that in the system water-cuticle-water, no asymmetry exists. For water-cuticle-air systems, when the membranes were dry-mounted they obtained asymmetry values of 2–3*x*, but if the membranes were not allowed to dry while being mounted high values averaging 10x but falling to 3–5x after a week were obtained. Asymmetry, they found, was independent of humidity, it was abolished by deep abrasion of the cuticle, but not (in contrast to Beament's results) by extraction with hot chloroform.

Several explanations have been advanced to account for this asymmetry of penetration. Hurst (1948) suggested a mechanism involving molecular or microvalves which has not met with general acceptance. Beament (1948) thought that evaporation might take place only through pores in the wax when the epicuticle is in air, but from the whole surface of the endocuticle when the cuticle is reversed, so that in the second position water would be sucked through the pores in the wax. Hartley (1948) showed that, in the general case, when the permeability of two different layers of a membrane is affected differently by the concentration of water within them, then the total permeability will differ according to which layer is next to water and therefore has the higher water concentration. This is a general statement and has not been shown to apply to the arthropod cuticle.

Beament (1954), in an important review of water transport in insects, has put forward a hypothesis which deserves close examination, particularly because a mechanism is proposed to account for asymmetry and also for the absorption of water vapour from unsaturated air.

Using a very thin membrane of tanned gelatin, Beament measured the rate of water uptake when immersed in liquid water and when hung in saturated air. The rate of loss of water when a fully saturated membrane was hung in dry air was also measured. These rates were graphed against water content of the gelatin expressed as a percentage of the saturated water content in liquid water (fig. 21). The rate of absorption from liquid water is very much greater than from saturated air, and falls off with rising water content until saturation is complete. Absorption from saturated air ceases when the water content is about 16 per cent of the liquid saturation value. To this extent the gelatin resembles a sponge: liquid water flows in rapidly by 'suction pressure', water vapour is absorbed slowly by a 'hygroscopic force'.

Now the rate of transfer across a tanned gelatin membrane

was found to be 15 mg./cm.²/hr. in standard conditions, and by reading off the curves it is possible to estimate the water content in the surface next to water (from curve *a*) and next to air (curve *c*), and thus to calculate that a concentration difference of 4·7 per cent between the two surfaces gives rise

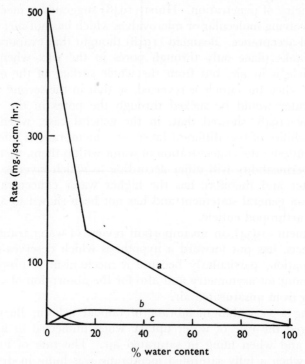

FIG. 21. *a*, Graph showing rate of uptake of a very thin tanned gelatin membrane when in liquid water; *b*, rate of loss of water when in dry air; *c*, rate of uptake of water when in 100% R.H. (From Beament, 1954.)

to a flow of 1 mg./cm.²/hr. (unit flow). The rate of transfer across a waxed gelatin membrane is also known—it is 2 or 8 mg./cm.²/hr. according to whether the gelatin or wax surface is against water. Knowing the concentration difference necessary to produce unit flow, it is possible to calculate the actual concentrations in the two sides of the gelatin necessary to account for the rate of flow in these conditions. The values for the two positions are shown in fig. 22.

In position (i), when the wax is in air, outward movement of water through the wax caused by evaporation at the surface is opposed by a suction pressure at the wax-gelatin interface because the gelatin is unsaturated and in contact with *liquid* water in the wax. In position (ii), the suction pressure at the wax-gelatin interface is very much greater, and there is no evaporation at the wax surface to oppose it. Flow in this position is therefore much faster than in position (i).

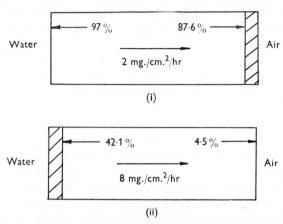

Water   97 %   87·6 %   Air

2 mg./cm.^{2}/hr

(i)

Water   42·1 %   4·5 %   Air

8 mg./cm.^{2}/hr

(ii)

Fig. 22. Beament's hypothesis for asymmetrical permeability of arthropod cuticles. Water concentrations (percentage of saturated values) and rates of flow in waxed gelatin membranes. Hatched areas, wax; plain areas, gelatin. (Data from Beament, 1954.)

It is a little difficult to imagine precisely what is happening in the wax layer of this model. It is essential for Beament's thesis that water in the wax shall be in the liquid phase, for suction pressure can only be exerted by gelatin on liquid water. But there is general agreement that the wax molecules at the inner surface are orientated and tightly packed. Can water molecules inside such small intermolecular spaces be in the liquid phase? More precisely, can the gelatin exert suction force upon such molecules as it does upon molecules of liquid water?

If, on the other hand, there are pores in an otherwise impermeable wax layer, the system is easier to envisage. The pores in position (ii) would be filled with liquid water, so that suction pressure could be exerted by the gelatin at their base. In position (i) the pores might be filled with air and water vapour, in which case the gelatin could not oppose evaporation by exerting suction pressure, but evaporation would be relatively slow, since it would take place only from the pores and not from the whole surface of the wax.

If pores or pinholes are permitted, the model can be made exceedingly simple. Consider any system composed of a layer of water-permeable material coated on one surface with a less permeable layer containing pores. When water is against the permeable layer, flow across the system will be limited by the rate of evaporation through the pores in the impermeable layer, and if the total area of these is small enough, evaporation will be less than from an uncovered surface. When the system is reversed, liquid water will penetrate the pores rapidly, and the impermeable layer will therefore offer no resistance to evaporation from the whole of the uncovered surface. This is similar to the system proposed by Beament in 1948.

This system will show asymmetrical penetration whether the permeable layer exerts suction force or not. If the pinholes are small, asymmetry will be large, if they are large, or if their number or size are increased by abrasion, asymmetry will be reduced and ultimately disappear. The system would only satisfy requirements if the wax in arthropods is at least partly superficial: if it is below the surface or dispersed through the epicuticle, asymmetry would not be expected, for the same area for evaporation would be exposed when the membrane is in either position. Again, if the wax is covered by a cement layer —which now looks like being a shellac (Beament, 1955)—the simple system would not work unless the cement too were porous or relatively permeable to water vapour. In *Tenebrio*, Holdgate has evidence that the cement layer is not superficial, neither is it in the cockroach (Beament, 1955), but it does appear to be so in *Rhodnius*.

Perhaps the most important question is whether the wax is discontinuous. When sense organs are present in the cuticle there is strong presumptive evidence that it is (Slifer, 1954, 1955). Transpiration/temperature relations, too, are not inconsistent with the occurrence of pin-holes in the wax, or indeed larger interruptions in some insects.

MECHANISM OF WATER-VAPOUR ABSORPTION. We may now return to consider again the mechanism of active absorption of water vapour from air. Beament (1954), developing the argument in the terms of his model described above, proceeds as follows: the rate of evaporation into air at 90 per cent R.H. from the waxed surface of a standard gelatin membrane is 2 mg./cm.^{2}/hr.; but this is also the rate of uptake by gelatin from liquid water when the gelatin is about 99 per cent saturated (from curve a in fig. 21). If, therefore, water at the wax-gelatin interface can be considered as free liquid water, it is only necessary to reduce the water content of the gelatin from the saturated condition to 99 per cent saturated in order to produce a suction force capable of resisting evaporation from the wax. Now in the epicuticle, the pore canals, filled with living matter, end in the tanned lipoprotein layer, perhaps only 1 μ from the inner surface of the wax. Surely, Beament argues, it is within the ability of living cells to regulate the water content of the tanned layer to this extent (1 per cent), and thus to take up water from humidities lower than that in equilibrium with the blood fluids.

There are two difficulties about accepting this thesis. First, there is no evidence that water in the molecular pores of the wax is either 'liquid' in the required sense, or 'free' in the sense of being free to move without hindrance by capillarity, etc. Secondly, if the suction force exerted on free water by 99 per cent saturated tanned protein is capable of preventing evaporation from its surface into 90 per cent R.H., then equally the force which ultimately removes the same water from the tanned protein must also be in equilibrium with 90 per cent R.H., and we are back at the beginning of the problem, i.e. the difficulty of imagining osmotic pressures in living cells of the order of 140 atmospheres, which are in equilibrium with 90 per cent R.H. Beament's thesis, so far as it applies to water-vapour absorption, must, therefore, be considered not proven.

WATER RELATIONS OF THE EGG STAGE. The shell or chorion of arthropod eggs is formed mainly by the follicular cells of the mother and differs from the integument of later stages in containing no chitin. Later in development the serosa lays down below the vitelline membrane the serosal cuticle or secondary egg membranes, which usually consist of an outer thin 'yellow' layer composed of a tanned lipoprotein like the cuticulin of the epicuticle, and an inner, thick, 'white' layer composed largely of chitin. The structure of the chorion itself is complex. It has been studied in *Rhodnius* by Beament (1946*b*).

In view of their very small size, the necessity for water-proofing eggs laid in dry surroundings is apparent. Beament (1946*a*) found that the whole of the chorion of *Rhodnius* is permeable to water, and that impermeability is conferred first by a primary wax layer, laid down by the oocyte itself on the inside of the chorion. Later, a secondary wax is laid down by the serosa, and subsequently permeates the secondary egg membranes (Beament, 1949). Primary wax layers below the chorion have also been demonstrated in *Melanoplus* (Slifer, 1946; Salt, 1952), and in *Lucilia* (L. Davies, 1948). In ticks there is a superficial wax layer secreted on to the outer surface of the egg by Gené's organ, and no primary wax layer (Lees and Beament, 1948). Red spider eggs possess both a superficial and a primary wax underneath the chorion. In *Locustana*, Matthée (1950, 1951) believes that a secondary wax layer is secreted via the pore canals in the white cuticle on to the yellow/white cuticle interface. According to Salt (1949, 1952) the secondary wax layer of *Melanoplus* is laid down by the serosal cells outside the secondary egg membranes.

These waxes show the same sort of transpiration/temperature curves as those of the epicuticle, except that permeability may be greater at a given temperature. Thus Beament (1946*a*) found the 'critical' temperature for the epicuticular wax of *Rhodnius* to be 57·5° C. and for the primary egg wax, 42·5° C. For the tick *Ornithodorus*, the corresponding figures were 62·5° C. and 45° C. In this tick, it is believed that the wax gradually permeates the shell from outside during incubation, because abrasive dusts and chloroforms have less effect in increasing evaporation as the egg gets older (Lees and Beament, 1948).

Changes in permeability of *Rhodnius* eggs have been studied in detail by Beament (1949), and correlated with the occurrence and distribution of water-proofing waxes. As mentioned above, the primary wax allows transpiration to increase rapidly in the region of 42° C. Just before blastokinesis, on the 6th day, transpiration drops considerably, and this coincides with the deposition of a second wax with a higher 'critical'

66

point. There is evidence that this wax permeates the whole of the secondary egg membranes; the egg shows a more gradually rising temperature/transpiration curve (fig. 23) and the wax is not removed by surface abrasion. But, as we have already seen, caution must be exercised in deducing cuticle structure from the shape of transpiration curves. Just before eclosion,

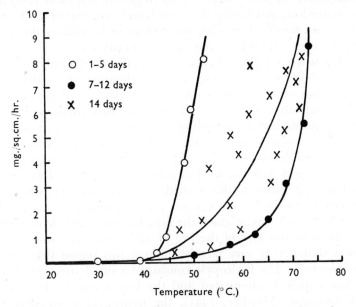

FIG. 23. Graph of the loss of weight of single eggs when desiccated at various temperatures, showing the sharp "transition point" in eggs 1–5 days (primary wax layer); the much smoother curve of eggs 7–12 days with transition at a much higher temperature (primary wax × secondary wax) and points (X) obtained with 14-day-old eggs when the secondary wax is being emulsified. (From Beament, 1949.)

the embryo secretes a wax emulsifier which attacks the secondary wax and transpiration becomes erratic (fig. 23), tending to rise towards the earlier level.

Owing no doubt to the presence of these waxes, some arthropod eggs are capable of developing in unsaturated air. *Lucilia* eggs at 37° C. develop satisfactorily at 50 per cent R.H. (L. Davies, 1948). In dry air these eggs lose water at the rate of 0·5 mg./cm.²/hr., which is low compared with many adult insects. *Rhodnius* eggs are even less permeable, and transpire

67

at 0·1 mg./cm.²/hr. in the same conditions. *Lucilia* eggs, however, require a higher humidity for hatching, as do many other muscid eggs, because rupture is assisted by differential absorption of water by the two layers of the chorion (L. Davies, 1950).

Eggs of the spider *Agelena* need at least a short exposure to humidities above 50 per cent each day if development is to be completed (S. Jones, 1941). Other eggs are still more resistant to water loss, and Leclercq (1948) found that eggs of the beetle *Melasoma populi* all develop successfully in humidities down to 17 per cent, and a few even survived in dry air.

The water relations of eggs are complicated not only because permeability to water may vary considerably during development, but also because many eggs, particularly those of acridiids and others which are laid in moist surroundings, absorb water during certain periods of development. Such absorption has been known for some time (Buxton, 1932a). Absorption from plant tissues in which the eggs are laid has been found in the capsid *Notostira* by Johnson (1937), and in *Trialeurodes* whose egg bears a thin-walled terminal bladder which is inserted into the leaves of plants (Weber, 1931).

Slifer (1938) investigated the area known as the 'hydropyle' at the posterior end of the egg of *Melanoplus differentialis*, and showed that it is very largely through this region that water is absorbed. The hydropyle consists of a small area where the white cuticle is attenuated and the yellow cuticle thickened. In *Locustana* (Matthée, 1951), the hydropyle is secreted by special hydropyle cells of the serosa. Cytoplasmic filaments from these cells apparently penetrate the fine pore canals in the white cuticle and link up with larger canals in the thickened outer yellow cuticle (fig. 24).

In *M. differentialis* the hydropyle is secreted by the serosal cells about the 6th day of development, and its appearance coincides with the onset of water absorption which continues until blastokinesis is complete and diapause sets in. If the hydropyle is sealed, no water is absorbed and no development occurs, but if water has already been taken up, sealing the hydropyle has no effect upon further development.

Salt (1949, 1952) found that the eggs of *Melanoplus bivittatus*, like those of other acridiids, must absorb water for successful development. Absorption occurs largely through the hydropyle

at the posterior pole of the egg. During the first few days very
little water is absorbed (2 per cent of the total weight in 5 days),
thereafter the serosal cells secrete the cuticula and the hydropyle, and water absorption increases rapidly, becoming greatest
at the 8th or 9th days and then decreasing owing to the limited
expansion of the cuticular layers. During this time, if eggs are

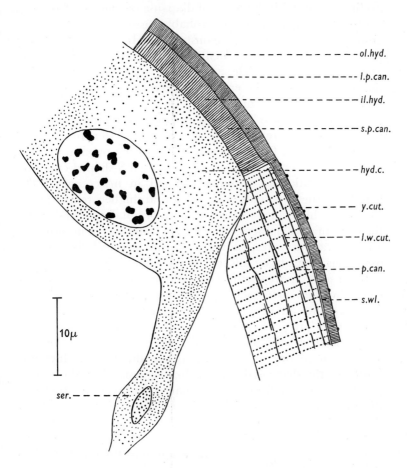

Fig. 24. Semidiagrammatic representation of the posterior tip of the egg of
Locustana pardalina showing a small portion of the hydropyle region and adjacent
unspecialized yellow and white cuticles; *hyd.c.*, hydropyle cell; *il.hyd.*, inner layer
of hydropyle; *l.p.can.*, large pore canals in outer layer of the hydropyle; *l.w.cut.*,
lamellae in white cuticle; *ol.hyd.*, outer layer of hydropyle; *p.can.*, pore canals in
white cuticle; *ser.*, serosa; *s.p.can.*, small pore canals in inner layer of hydropyle;
s.w.l., secondary wax layer; *y.cut.*, yellow cuticle. (From Matthée, 1951.)

kept in dry air, loss of water by transpiration occurs: it is kept low in young eggs by the primary wax layer, and later by the secondary wax layer above the cuticula. But from the 8th to the 14th day (at 25° C.) transpiration is very greatly increased, and occurs largely through the hydropyle. After blastokinesis (about the 14th day) the desiccation rate is very variable, but declines again with the onset of diapause on the 21st day.

Other specialized areas probably associated with water absorption have been demonstrated. In the plecopteran *Pteronarcys* Miller (1940) demonstrated a thickened circular area in the serosal cuticle, and Tiegs (1942a, b) considers that the dorsal organ of the embryos of certain Apterygota is concerned with water absorption.

In *Gryllus commodus* T. Browning (1953) found no specialized water-absorbing area, but the eggs took up water from a moist surface, slowly at first, then rapidly, until the maximum water content had been reached, giving the usual S-shaped curve for water absorption so commonly found in insect eggs. The length of these periods was shorter at higher temperatures (fig. 25), which suggests that the rate of uptake depends upon the stage of development. The rate of water loss in unsaturated air also varied correspondingly, so that permeability in both directions seems to be affected at the same time.

Birch and Andrewartha (1942) found that rapid uptake of water by *Austroicetes* eggs was postponed till after diapause; similarly, the rate of transpiration increased after diapause.

Laughlin (1953) found a similar state of affairs in the eggs of the beetle *Phyllopertha horticola*, where water absorption and desiccation in dry air both increased after a few days and subsequently decreased. Laughlin suggests that the waterproof layer of the fresh egg (primary wax layer) is broken by the embryo during the period of water absorption and restored thereafter.

Laughlin believes water absorption in *Phyllopertha* to be osmotic. This, however, does not seem to be the usual mechanism, for in *Melanoplus* the rise in rate of absorption with temperature gives a Q_{10} value of 10 or 12, whereas if the action were simply osmotic, rates would be proportional to the absolute temperature. Furthermore, Matthée (1951) has found that aerobic respiration is necessary for water uptake in eggs of

Locustana pardalina, and that these eggs can absorb water from hypertonic glucose solutions (1·75 per cent NaCl equivalent). In *Melanoplus* uptake varies with metabolic activity irrespective of the membranes present (Thompson and Bodine, 1936). For these reasons, water uptake by eggs is generally thought to be an active process.

The evidence so far considered shows that in most eggs which absorb water this is restricted to a given period of development—it may be before diapause as in *Melanoplus,* or

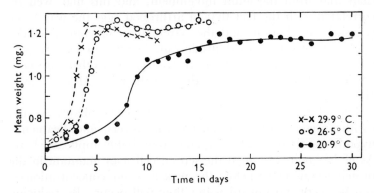

Fig. 25. Changes in the mean weight of eggs during development at three constant temperatures. Increase in weight is due to absorption of water. (From T. Browning, 1953.)

after as in *Austroicetes* and probably *Gryllus,* or it may occur in the middle of development when there is no diapause (*Phyllopertha*). But in all these eggs the period of absorption is also a period of increased transpiration if the eggs are put in dry air. In some cases water movement is controlled by the hydropyle, but in others (*Gryllus, Phyllopertha*) it appears to take place over the whole egg. Furthermore, changes in rate of permeability are not entirely accounted for by the secretion of wax layers, for changes occur after these have been laid down.

There are two further points to be noted: first, that absorption from unsaturated air has never been demonstrated (though absorption from hypertonic solutions seems to be possible in *Locustana*); secondly, it is important that eggs shall be capable not only of restraining water loss but also of resisting the inflow of excess water (which would occur if the process were a purely osmotic one). The eggs of *Gryllus* will develop and hatch entirely submerged in distilled water (T. Browning, quoted by Beament, 1954).

There are, then, two main water problems, so far as eggs are concerned: for those laid in dry surroundings the prevention of desiccation, for those laid in contact with liquid water, prevention of flooding with provision for absorption of water during a limited period. The first problem is solved by waxing. The absolute rate of loss per unit area from many eggs is greater than that from the adult integument, and this may well be correlated with the fact that in arthropod eggs the wax layer is inside the chorion, and the system therefore resembles the adult integument turned inside out, leading to greater permeability as we have seen above (p. 64). But others are very efficiently water-proofed, e.g. those of the mite, *Metatetranychus* (Beament, 1951).

The solution of the second problem is very probably assisted by the presence of impermeable wax layers, and Beament's hypothesis, discussed above (p. 63), is probably applicable here. As he has pointed out, if the serosal cells can maintain in the proteinaceous layer below the wax a water concentration of 100 per cent, no inward water flow will occur. By reducing the water content to 99 per cent, water uptake will be allowed as happens in the mid-embryonic period.

So far as absorption from hypertonic solutions is concerned, this has only been demonstrated in *Locustana* and may be rare. The osmotic barrier is much less formidable than when water vapour is absorbed, but the process is clearly an active one.

Thus the biological activity of living material is again seen to be involved in problems of water transport. The presence of wax layers, by offering resistance to the passage of water, may reduce the problem to manageable proportions, but the establishment of the necessary gradients either to facilitate or prevent flow must rest with the activity of the living material, and on that side of the problem very much remains unexplained.

CHAPTER 5

WATER AND BODY TEMPERATURE

<hr>

ALL terrestrial arthropods lose a certain amount of water by transpiration in dry air, and in some species the rate of loss is rather rapid. It is to be expected that transpiration will depress the body temperature, but whether or not such depression is significant depends upon the size of the contribution made by evaporation as compared with other factors in the total heat balance. The factors concerned are radiation, conduction, convection, metabolism and evaporation; and the equilibrium temperature reached is that at which total heat lost is balanced by total heat gained. Each of the factors mentioned seems to depend, for a given species, more upon surface area than upon volume, so that the final equilibrium temperature reached will be independent of size, though the time taken to reach equilibrium will of course be larger for larger animals.

Earlier work on this subject has been reviewed by Gunn (1942). Up to that time, temperature depressions (from ambient) had been recorded in a number of insects: by Necheles (1924) and Mellanby (1932c) for *Periplaneta*, Bodenheimer and Samburski (1930) for *Schistocerca*, Koidsumi (1935) for *Gastrimargus* and others, and it was clear that such depressions, which amounted to 5° C. in some cases, only occurred in unsaturated air. These depressions were recorded in laboratory conditions, where radiation, conduction and convection were negligible.

TRANSPIRATION AND UPPER LETHAL TEMPERATURE. If transpiration has a significant effect upon the body temperature of arthropods, they should be able to withstand measurably higher ambient temperatures in dry than in moist air.

Mellanby (1932c) exposed several species of insects for periods of 1 hr. and 24 hr. to different humidities, and recorded the highest temperatures in which they survived (fig. 26, full

73

lines). For 1 hr. exposures he found that the lethal temperature for any one species was the same at all humidities. Mellanby explained these results by supposing that at low humidities, excessive evaporation caused death by desiccation so that no higher temperatures could be withstood. The larvae of *Tenebrio*

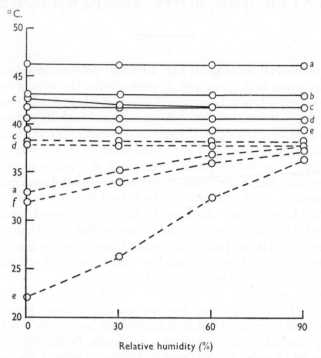

FIG. 26. Highest temperatures at which certain insects can withstand exposure at various relative humidities for 1 hr. (full lines) and 24 hr. (broken lines). *a, Pediculus; b, Lucilia; c, Tenebrio* larvae under 30 mg.; *c', Tenebrio* larva over 100 mg.; *d, Xenopsylla* adults; *e, Xenopsylla* larvae; *f, Anopheles.* (After Mellanby, 1932*c*.)

were exceptional, for if they were above a certain size, they could afford to lose water at a high enough rate per unit area effectively to reduce their body temperature below ambient for this period of time.

This interpretation is in my view open to question. It may be assumed that the highest temperature an insect can withstand at 90 per cent R.H. is close to its true thermal death point, for very little evaporation can take place. In dry air, according

74

to the accepted explanation, evaporation occurs and leads to a just tolerable loss of water. It also leads to a temperature depression such that the highest air temperature which can be withstood is the same as that which can be withstood at 90 per cent R.H. (i.e. its true lethal temperature). Now there seems no reason why that rate of evaporation which is just not lethal after 1 hr. should also reduce the body temperature by such an amount that the insect can withstand an air temperature precisely equal to its thermal death point. By coincidence this might occur in one set of conditions, but when it is supposed to happen to all the insects used and at all humidities, credulity is strained.

A simpler interpretation is that so little evaporation took place that the slight rise in tolerable air temperature went unnoticed. In other words, the true lethal temperature (within the limits of experimental error) was being measured at all humidities, and this does not vary. In the large *Tenebrio* larvae evaporation was apparently rapid enough to produce a measurable temperature depression. Alternatively, because of their large size and consequent large thermal capacity, these larvae took longer to heat up than the smaller insects, and were not at the temperature recorded for the whole exposure.

If this is the correct interpretation, then no conclusion whatever can be drawn from the results as regards temperature depression by evaporation, and there need be no question of the insects just not succumbing to desiccation.

For long exposures, however (fig. 26, broken lines), the effect of low humidities becomes very pronounced for those insects which are not efficiently water-proofed. Consequently they can only withstand much lower temperatures in dry air than in moist—not because of the effect of temperature, but because in any higher temperature evaporation would rise and lead to death by desiccation.

More direct evidence of temperature depression was obtained by Gunn and Notley (1936), who found that three species of cockroach could withstand higher temperatures in dry than in moist air, during short exposures. In 24 hr. exposures the effect was reversed. Beattie (1928) and Smart (1935) found essentially similar results with blow-flies and cheese skippers respectively.

In woodlice, transpiration is in general much higher, and the effects on body temperature more pronounced. Edney (1951*b*) measured the body temperature of various species in dry and in moist air, and obtained depressions varying from less than 1° C. to as much as 7° C., according to the species and conditions used (Table 3). The insect *Blatta orientalis* was also used in this work. It is approximately the same size and shape as the oniscoid *Ligia oceanica*, but it loses water less rapidly, and its body temperature depression was also found

TABLE 3. *Mean depression of body temperature of woodlice and* Blatta orientalis *below that of the surrounding air, after 30 min. exposure*

(Data from Edney, 1951*b*.)

	Air temperature °C.	Depression °C.
Ligia oceanica	20	2·6
	37	6·8
Onistus asellus	20	1·5
	37	2·7
Armadillidium vulgare	20	0·5
	37	1·8
Porcellio scaber	20	0·4
	37	1·3
Blatta orientalis	20	0·7
	37	2·4

to be less. The upper lethal temperatures for woodlice have also been determined in various conditions (Edney, 1951*a*) (fig. 27). For 15 min. exposures the upper lethal temperature is clearly higher in air at 50 per cent R.H. than in saturated air. In perfectly dry air there is no further rise, sometimes even a fall, in the lethal temperature, from which it appears that harmful effects of desiccation begin to outweigh the beneficial effects of cooling. For 24 hr. exposures, the effects of desiccation in unsaturated air greatly outweigh the advantages, and lethal temperatures are therefore much lower in dry air. It is interesting to see that *Ligia* behaves differently from the other oniscoids in showing a higher apparent lethal temperature at 0 per cent than at 50 per cent R.H. in short exposures. This is undoubtedly a result of its much larger size, so that it can afford to evaporate more water per unit area, and thus to cool itself for longer, than the rest.

The general impression derived from this field of work is that even in laboratory conditions but little temperature depression below ambient can be expected in arthropods which are efficiently water-proofed (except perhaps for a rather short time by the evaporation of surface water, as suggested by

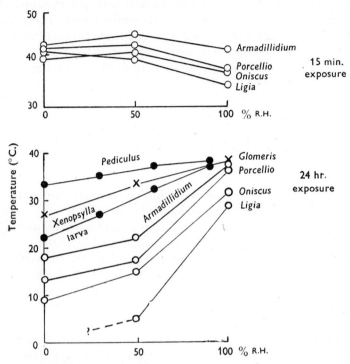

Fig. 27. The highest temperatures which various woodlice, *Glomeris* and certain insects can withstand for 15 min. and 24 hr. exposures at various humidities. (From Edney, 1951*a*.) Data for *Xenopyslla* and *Pediculus* from Mellanby, 1932*c*.)

Waterhouse (1951)). In those where the natural rate of transpiration is higher, such as cockroaches and woodlice, significant depressions occur, but it is only in larger species that the necessary rate can be maintained for long. The extent to which such temperature depression is significant in natural conditions must now be considered.

TRANSPIRATION AND TEMPERATURE IN THE FIELD. We have seen that transpiration may significantly affect the body temperature of arthropods in laboratory conditions, and it has often been assumed (Buxton, 1924a), perhaps too easily, that transpiration serves to reduce their body temperature in natural conditions, when they would otherwise become dangerously hot. Such temperature regulation occurs of course in mammals, but these are large animals with a comparatively small surface area, and they can therefore afford to transpire at a high rate per unit area for a considerable period of time.

The most successful terrestrial arthropods have a relatively impermeable integument, so that apart from the danger of losing too much water, there is also the question as to whether they can transpire fast enough to effect a significant temperature reduction.

The problem is of course most acute in extreme terrestrial conditions such as deserts where humidity is low and temperature high. Buxton (1924a, b) was one of the first to point out the problems facing small animals living in such places, and he also pointed to one of the solutions, namely the existence of very great differences in microclimate which exist within small areas, and which may be taken advantage of by animals. The point has been confirmed and elaborated by many authors including Bodenheimer (1935) and particularly by Williams (1923, 1924a, b) in his work on the bioclimate of the Egyptian desert. But there is still a lack of quantitative information about the interplay of climatic factors in determining body temperatures in natural conditions.

Measurements of body temperature have not infrequently been made (references in Gunn, 1942; Uvarov, 1948) showing the effect of solar radiation in raising the body temperature. In general the animals studied appear to use this source of heat to bring their body temperature up to optimum level. But there was until recently little or no information on rates of transpiration, or on the effects of transpiration, in the field.

In order to measure the effect of evaporation alone in such conditions a completely dry animal can be exposed side by side with a normal living one, and the difference in temperature noted. At the same time heat transfer from other sources may be estimated and the relative significance of evaporation assessed.

D. Parry (1951) came to the conclusion on theoretical grounds that in direct sunlight, transpiration plays a very insignificant part in determining the equilibrium temperature of arthropods, and Digby (1955) reached the same conclusion for small insects even in the absence of radiation. Parry expressed all the heat exchanges involved in terms of mW./cm.^{2} and was thus able to make a comparison between each. Evaporation of 1 mg. water/hr. involves a heat consumption of 0·67 mW. Calculations based on Ramsay's (1935b) data for *Periplaneta* lead to an expected evaporation rate of about 1·7 mg./cm.^{2}/hr. into briskly moving air at 20° C. when sunlight has warmed the animal to 35° C. This is equivalent to a heat loss of about 1·1 mW./cm.^{2}/hr. The net radiation load (input less output) upon a cockroach when the sun's altitude is 45° is something between 10 and 50 mW./cm.^{2}/hr., so that evaporation would hardly affect the heat balance in such conditions. Metabolism makes a contribution of less than 0·5 mW./cm.^{2}/hr. Convection of heat away from the animal plays a large part in the heat balance—it varies according to wind speed and temperature difference (Digby, 1955). Normally the temperature of the animal will rise as a result of radiation, thus causing an increase in the heat lost by convection, until the total heat lost balances that gained. Conduction, in normal circumstances, when an animal is supported on the tarsi of its legs, is very small.

But not all arthropods have such impermeable cuticles; even insects at high temperatures, or insects with the spiracles fully open, may be expected to evaporate more water, and in such cases evaporation might play a significant part in temperature reduction. Edney (1952, 1953) has found that transpiration may indeed play a significant part in lowering the body temperature of woodlice exposed to the sun. Measurements of the rectal temperature were made of living, freshly killed and dead, dry specimens of *Ligia* exposed to solar radiation upon a surface at about 35° C. when the air temperature 2 cm. above the surface was 21° C. In these circumstances the temperature of both the living and the freshly killed specimens dropped to 28° C. while that of the dry animal remained within 0·5° C. of the surface temperature. This temperature depression of 7° C. must therefore have been caused by evaporation, and an estimated heat balance-sheet in terms of mW./cm.^{2}/hr. read as follows:

15 (radiation) + 2·5 (conduction)
$$= 9·8 \text{ (convection)} + 7·7 \text{ (evaporation)}.$$

Comparative measurements of temperature depression in several species of woodlice and in the cockroach *Blatta* always showed the temperature of the insect to be higher than that of *Ligia* (fig. 28), though the two are approximately the same size and shape, while the temperatures of the other species of woodlice were depressed, with some exceptions, according to their relative rates of transpiration.

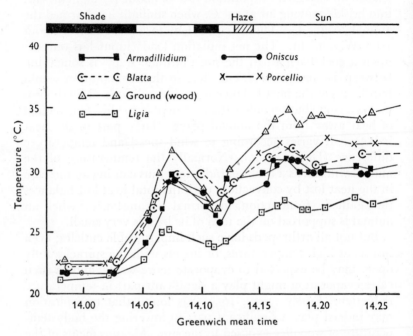

FIG. 28. The internal temperatures of various species of woodlice and the cockroach *Blatta*, exposed simultaneously to insolation, upon wood (light deal). Air temperature 5 mm. above the ground followed that of the living *Ligia* closely. 14 August 1952. R.H. 55–63%; wind speed *c.* 50–100 cm./sec. (From Edney, 1953.)

Woodlice, though normally cryptozoic like many other arthropods, are in nature frequently to be seen in direct sunlight. Usually this occurs when the effect of such insolation is to raise the temperature nearer to the optimum (Edney, 1953, and unpublished), but occasionally, if they are trapped in a very humid microclimate which also becomes hot as a result of insolation they may be forced to emerge into the sunshine where, owing to evaporation and convection, the body temperature falls (Edney, 1953) (fig. 29). In such ecological crises as this, the ability to transpire rapidly is probably of survival value by preventing too great a rise in temperature.

WATER AND RESISTANCE TO LOW TEMPERATURES. Many insects from normally warm surroundings are unable to survive low temperatures above freezing point—honey-bees,

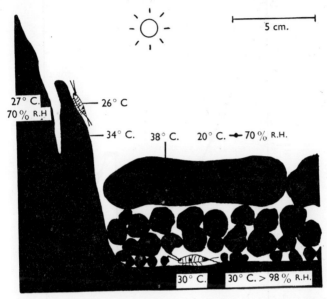

FIG. 29. Vertical section of base of red sandstone cliff and shingle inhabited by *Ligia* (diagrammatic). To show the microclimatic conditions and internal temperatures of the animals, at *c*. 14.00 G.M.T. in August 1951. (From Edney, 1953.)

for example, die between 1° and 8° C. A few insects are capable of surviving after their tissues are frozen (Payne, 1927*a*; Salt, 1936), but the majority of insects only survive temperatures below freezing point provided their tissues do not freeze. Survival in these cases is due to the phenomenon of supercooling, whereby water in the body behaves like water in a capillary tube and ice does not form until the temperature has been reduced far below 0° C. This property in insects has been known since 1899 when Bachmetjew published his paper on the body temperatures of insects. His figure showing the fall in temperature of *Saturnia pyri* is well known. Exposed to an air temperature of $-13.5°$ C. the body temperature fell slowly to $-9°$ C. when freezing occurred (the undercooling point), rose suddenly to about $-1.5°$ C. (the rebound point) owing to liberation of the latent heat of freezing, and then commenced

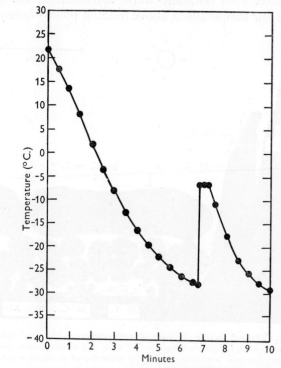

Fig. 30. Typical cooling curve for the codling moth larva, showing under cooling and rebound temperatures. Air temperature: $-37°$C. (From Siegler, 1946.)

to fall again. His results have subsequently been confirmed by a number of workers with different insects and similar temperature curves obtained (e.g. Siegler, 1946) (fig. 30). Hardiness in most insects extends as far as the undercooling point, and this varies considerably between species. If the tissues are encouraged to freeze by pricking, the true freezing point is obtained, and this is about $-1°$ C. to $-2°$ C. for most insects. There is a good deal of evidence that cold hardiness increases during the winter months in those insects which hibernate, such as *Chortophaga* (Bodine, 1921) and the moths, *Isia isabella* and *Diacrisia virginica* (Payne, 1927*b*). In other insects

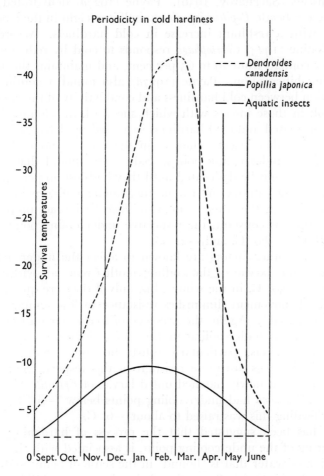

FIG. 31. Periodicity in cold hardiness. *Dendroides canadensis* – – –, *Popillia japonica* ——, aquatic insects — —. (From Payne, 1927*a*.)

not exposed to extreme cold, such as *Popillia japonica*, which hibernates below the frost level in the soil, and aquatic insects, annual cycles of cold hardiness are not so marked, or are absent (fig. 31).

83

It also happens that the water content of insects is often less during hibernation than at other times, and this has led to the belief that reduction of the water content causes hardening in insects (Saccharov, 1930). Payne (1927a) dehydrated the Japanese beetle *Popillia japonica* to half its weight in the laboratory with a resultant increase in cold hardiness. According to Bodine (1923) *Chortophaga* responds to cold by reducing its water content from 79 to 65 per cent, and maintains this level during hibernation. Payne (1927a) also found a continuous relation between cold hardiness and conductivity of the haemolymph in those insects with which she worked. Conductivity was decreased and cold hardiness increased by dehydration.

Yet in other insects, changes in the water content do not affect the undercooling points. Salt (1936) found that adult *Leptocoris* could be desiccated until they had lost 20 per cent of their weight without effect upon their undercooling point, and that although the water content of *Ephestia* larvae, prepupae and pupae is constant, the respective undercooling points are $-5 \cdot 8°$ C., $-8 \cdot 0°$ C. and $-21°$ C.

Several other factors are known to affect the undercooling points; thus larvae of the codling moth *Cydia* can withstand -24 to $-30°$ C. in the winter, but only if they are quiescent; artificial movement diminishes resistance to low temperature (Siegler, 1946). Again, the presence of free water on the surface prevents undercooling in *Leptocoris* (Hodson, 1937), and in other insects, particularly soft-bodied ones (Salt, 1936), though the eggs are more resistant in this respect. Salt also observed that newly hatched unfed larvae of *Ephestia*, *Sitotroga* and *Malacosoma* had undercooling points below $-20°$ C., but after feeding this was raised to about $-6°$ C.

It has been suggested that the process of hardening, i.e. lowering of the undercooling point, is associated with the proportion of water that is 'bound' in the sense of being unfreezable. Robinson (1928a, b) found that in pupae of *Telea polyphemus* and *Callosamia promethea* subjected to gradually falling temperatures to $-14°$ C., the 'bound' water rose from 13 to 40 per cent, while the 'bound' water in other non-hardy species did not increase. But the nature and physical properties of 'bound' water have always been a little mysterious. Certainly there is always a residuum of unfreezable water, even

84

at $-40°$ C. Ditman *et al.* (1942) found that the percentage of unfrozen water for any one species varied considerably between the true freezing point and the undercooling point (fig. 32), which may be due to the fact that the insects were measured in batches, perhaps containing variable proportions of frozen and

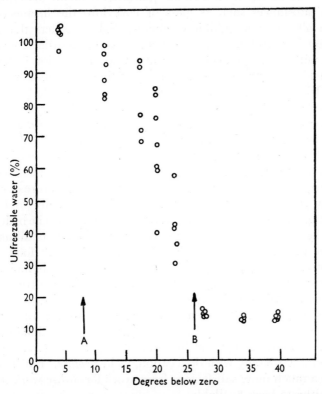

FIG. 32. Percentage of total water unfrozen at various subzero temperatures in cold-hardy overwintering European corn-borer larvae. Determinations made in February. Each point represents a single sample of 15 insects. (From Ditman *et al.*, 1942.)

unfrozen individuals. Nevertheless, below the undercooling point the unfrozen water content was fairly constant for each species, and, more important, bore no relation to the undercooling point itself. The percentage of unfreezable, or 'bound' water cannot therefore be said to be the controlling factor in determining hardiness.

Salt (1953) has recently published an important paper on the effects of food and other foreign material upon undercooling points. He found that larvae of *Ephestia* and of *Agrotis* go through a cycle of hardiness which corresponds to their feeding periods; they are much less hardy when food is in the gut. The inoculation of small quantities of plant material or inert substances such as glass or cork into the haemocoele, were all found to inhibit hardiness. Salt considers that the correct interpretation of a sudden increase of hardiness associated with a change of stage or entry into hibernation is that a reversion occurs to the natural hardiness peculiar to the insects' tissues from a reduced hardiness caused by the presence of food in the gut acting either as a more easily freezable substance, or simply as a foreign body, but in either case leading to general freezing by inoculation of ice crystals.

Salt contends that the process of gradual hardening is unproven. When this has been reported, as by Decker and André (1936), the results may have been caused by a mixture of cold hardy and non-hardy individuals, one kind gradually predominating. Evidence concerning the effects of water loss he believes may be attributable to cessation of feeding (before moulting or before hibernation) leading to a reduction of moist food and digestive juices in the gut. To show that a decrease in moisture content *causes* hardening needs more than a mere correlation.

This work has undoubtedly thrown fresh light on problems of cold hardiness. It does not succeed in accounting for all the cases where true hardening is associated with reduction in water content as reported for example by Payne (1927a), or artificial hardening with desiccation, as in *Popillia*. But it provides a much more satisfactory framework for future work than has hitherto been available.

The essential difference between Salt's hypothesis and that of earlier workers is that he believes hardiness of the *tissues* to be unchangeable: hardiness of the whole insect being determined by the presence of foreign bodies. Other workers, notably Payne, believe that hardiness of the tissues may be altered by changes in water content, or concentration of watery solutions present.

Perhaps both foreign bodies and changes in water content are effective, but to different extents in different insects.

86

CHAPTER 6

GENERAL CONCLUSIONS

EVEN within the limits selected somewhat arbitrarily for the purposes of the present monograph, the study of water relations has proved to include a number of rather diverse fields of inquiry. This is perhaps no more than a reflexion of the importance of water in so many biological processes. However, we are now in a position to take a rather broader view of the subject and to attempt to establish relations between the various lines of work and the central theme, which concerns the nature and extent of the problems involved in maintaining a satisfactory water balance, and the means which have been adopted for their solution.

We have seen that because of their small size, terrestrial arthropods are particularly vulnerable in dry surroundings. Evidence of this is the comparative lack of success, as measured by numbers of species, or by the number of different habitats occupied, of those groups of arthropods which are less well adapted to meet the problems of water economy, and which are thereby confined to a cryptozoic existence.

There are three main potential sources of water loss: excretion, transpiration from the integument, and transpiration from the respiratory surfaces. It is possible almost entirely to eliminate loss by excretion by the use of low-hydrogen, insoluble compounds of nitrogen which can be excreted dry, and this is done in the more successful groups, insects and arachnomorphs, which excrete uric acid and guanin respectively. But uric acid is also found in centipedes, which are not particularly well adapted to dry conditions, and also in the Onycophora, which are certainly not. There is, therefore, no clear correlation between a terrestrial habitat and the evolution of uricotely. The woodlice are also aberrant in this respect, for although some species inhabit less moist surroundings than others, all are ammonotelic. But it would be useful to know more about excretion in these animals; the amount of water loss associated with the process is quite unknown.

It is also possible to eliminate tr anspiration from the integument to a large extent by water-proofing with wax. Here again, insects and arachnomorphs are conspicuously successful, and on the whole there is a good correlation between the physical properties of the wax and the habitat of the species— those from hotter, drier places have waxes which are less permeable, and remain so to higher temperatures. Other groups of arthropods have on the whole more permeable integuments, and again there is a general correspondence between permeability and degree of dryness of the habitat. This is true even within a group such as the woodlice.

Loss of water from the respiratory surfaces, however, cannot be eliminated, it can only be reduced by invagination and by closing devices operated by carbon dioxide concentration, so that they remain open no longer than is necessary. *Peripatus* and the millipedes have short bunches of non-anastomosing tracheae, and their spiracles are unprotected. There is no information about water loss from them; it is probably large, but in the absence of an impermeable cuticle there is little point in greatly reducing loss from the spiracles. In woodlice, too, respiratory surfaces are by no means well protected from water loss. In the more terrestrial forms there are indeed internal respiratory organs, but their openings are unprotected, and the amount of water lost is still considerable. The proportion of water lost from the respiratory surface is, of course, no guide to the efficiency of the protecting devices. Some 60 per cent of the total is lost through the spiracles of insects, and a much smaller proportion from the pleopods of woodlice, yet the absolute amount lost by insects is small—it forms a large proportion of the whole because of efficient water-proofing of the integument. The figures quoted nevertheless reflect the fact that it is easier to eliminate loss from the integument than from the respiratory surfaces.

In assessing the effects of various physical factors on water loss we have seen that temperature, humidity and wind speed are the most important. Earlier attempts to express the combined effects of temperature and humidity in one measure, saturation deficit, were only partially successful, not only for the theoretical reason that saturation deficit is not proportional to evaporation even from an inanimate body, but also because temperature certainly, and humidity probably, themselves affect the permeability of the cuticle as well as the evaporating power of the air. It is the departures from the 'saturation deficit law' rather than the law itself that have thrown light upon water loss from arthropods as living systems.

One of these departures from expectation, namely the excessive rise of transpiration from the integument at high temperatures, led to the discovery of the epicuticular waxes associated with water-proofing. We have seen that the earlier picture of a continuous, superficial layer of wax which undergoes a phase transition at a 'critical' temperature, must be re-examined, though the true position is by no means clear.

The earlier temperature/transpiration curves, although often interpreted as meaning a discontinuous sudden increase in permeability, in fact pointed to a continuous rise in permeability commencing rather abruptly at one temperature. But if allowance is made for the known rise in saturation deficit and thus of the drying power of air as temperature rises, and also for the known increase in the rate of diffusion, then the shape of the temperature/transpiration curves for terrestrial insects shows a fairly good exponential rise with temperature, while that for woodlice and aquatic insects shows no change in permeability with temperature.

The presence of a 'critical' temperature can no longer be used as an indication of the presence of a wax layer, and indeed lipoids have been extracted from arthropods which do not show a rise in permeability with temperature, though the distribution of these lipoids is doubtful.

The questions which need answering relate to the location and structure of the wax layers. It seems unlikely that a discrete epicuticular layer of wax is present in all arthropods. There is no reason to suppose that one exists in woodlice or millipedes, for example. The evidence for a superficial discrete layer in certain insects is, however, very convincing. A wax can be extracted from exuviae which, when deposited on inert membranes, confers properties of impermeability similar to those of insect cuticles; *superficial* abrasion leads to a loss of impermeability; and finally, although such a layer has never been demonstrated histologically (perhaps because of its extreme thinness), Holdgate and Seal (1955), by means of the electron microscope, have observed a superficial layer in *Tenebrio* pupae which appears to be soluble in wax solvents.

On the other hand, there is some evidence, though not in my view convincing, that wax permeates the whole of the serosal membranes in certain eggs. The recently reported fact (Beament, 1955) that the short-chain paraffin and alcohol solvents of the longer-chain waxes confer upon the latter remarkable spreading powers over water-saturated protein surfaces, also opens the possibility of penetration *into* such material; and besides enabling us to see how repair of a damaged wax layer through an intact cuticle may occur, also makes it easier to imagine penetration of the wax as reported in certain egg membranes, and opens the question as to whether, in some cuticles, wax may be present throughout the cuticulin layer of the epicuticle either instead of, or as well as, in a superficial layer. It would be interesting if the lipoid already known to be present in the cuticulin layer turned out to have similar properties to the superficial wax.

But even if the wax is superficial, the question still remains as to whether it is continuous. Asymmetrical permeability is most simply accounted for by a superficial, impermeable but porous layer, and an hypothesis to account for the salient features of various temperature/transpiration curves in terms of degrees of porosity has already been put forward (p. 26). But there is no direct evidence that it is porous, and the concept of a tightly packed, orientated layer of molecules at the wax-cuticulin interface, which is at least consistent with certain permeability and other phenomena, is a little difficult to associate with porosity.

A further, closely related question is the extent to which evaporation occurs through the wax, and whether the exponential temperature effect (p. 21) is a property of the wax or of other regions of the epicuticle. On the whole, the evidence points to evaporation through a wax layer, subject to an exponential temperature effect, as well as through non-waxed pores; the contribution of the former component being greater in less permeable cuticles where the pores are smaller. But the evidence itself is partly contradictory, and a radically new approach to the problem is much to be desired.

On the uptake side of the balance we have seen that absorption with the food or by drinking is commonly found in arthropods. The absorption of water from moist surfaces is a phenomenon which deserves further investigation. It has been demonstrated in woodlice and spiders, where it appears to be very useful, but the extent of its occurrence throughout the group is unknown.

Metabolic water is undoubtedly of value to animals in dry surroundings. It is indeed practically the only source of water in some cases. But one can hardly speak of the *use* of metabolic water as an adaptation to dry conditions—the adaptation consists in conserving water whatever its origin. The interesting question is whether dry conditions stimulate extra metabolism of food reserves. In some insects this does seem to occur, and it is then proper to speak of an adaptation to dry conditions. In a few insects, notably *Tenebrio* larvae, extra metabolism of dry materials is stimulated in low humidities to such an extent that maintenance of a constant wet/dry weight ratio is possible. Insects so well adapted to dry surroundings as this, however, are unable to regulate in moist air.

The absorption of water vapour from unsaturated air remains unexplained. It has been demonstrated conclusively only in a small number of arthropods, and this is a little surprising in view of the obvious value of the faculty for dry-living animals. But it may well be, as suggested earlier (p. 58), that we are dealing here with the extreme form of a much more widespread physiological process whereby activity of the epidermal cells restrains the passage of water outwards. The manner in which this would be accomplished is obscure. Beament suggests that the wax plays an important part in many processes of water transport, both inward and outward, by providing an inert system containing liquid water and reducing the magnitude of the osmotic problems. Although not entirely convincing so far as absorption of water vapour is concerned, this thesis may well be applicable in other cases, particularly in the regulation of water flow through the shell of eggs. A report which has not received much attention but which may be very relevant to the present discussion is that of

Govaerts and Leclercq (1946) who found that if insects are kept in air saturated with water vapour containing 8 per cent of heavy water, their own body water also contains 8 per cent heavy water after a period of 5–13 days, depending on the species. This, if confirmed, implies a continuous exchange between body water and external water. Loss or gain would thus represent the sum of inward and outward flow, and this might be positive or negative. Such an interpretation would lend support to the idea of inward secretion by the epidermal cells reducing net outward flow.

So far we have been dealing mainly with the prevention of water loss, and the means available for replenishment. The interaction of gain and loss must be so regulated as to maintain a constant internal environment, and this raises the question of internal osmotic and ionic regulatory mechanisms. Interesting advances have recently been made in this field as a result of the application of modern techniques, and we are now beginning to see how the Malpighian-tube and rectal system functions in a way analogous to the vertebrate glomerular kidney. Some arthropods possess remarkable powers of regulation based largely on the ability of the Malpighian-tube and rectal system (insects), or the coxal glands (argassid ticks), to excrete a urine whose osmotic and ionic properties vary according to need. On the other hand, most arthropods so far studied seem to be more tolerant of osmotic and ionic changes than mammals are, and in some, such as woodlice and ixodid ticks, osmotic tolerance within limits rather than regulation appears to be the rule.

It has been necessary to consider certain aspects of the temperature relations of arthropods, where these are clearly involved in water economy. But it is abundantly clear that although osmotic and ionic regulation are reasonably well established, there is little or no regulation of body temperature, except so far as choice of immediate environment may be considered as regulation.

It is true that among the less advanced or less terrestrial forms, transpiration is rapid enough to reduce body temperature in ecological crises for short periods, and indeed may have been of survival value during the transition from marine to land life for those animals which emerged via the littoral zone;

but it is only by the development of a high degree of temperature tolerance that small animals can take advantage of the fully terrestrial habitat. Impermeability of the cuticle, which is essential for this advance, may only be feasible when the twin problems of temperature tolerance and an efficient internal respiratory system have been solved.

Ability to regulate the internal environment in the face of external change is not a bad measure of biological success. Ability to regulate water and ionic concentration is well developed in some arthropods, but ability to regulate temperature is not, perhaps because of their small size. Consequently, for an arthropod, the necessity to find a satisfactory environment is much more pressing than it is for a mammal. The arthropod organization, particularly as exemplified by insects and spiders, is a good one, for it is capable of a great variety of modifications. This capacity has been exploited by evolving a large number of different species, each adapted to a rather narrow range of conditions. Thus the individuals of any one species are not faced with the necessity to regulate against a wide range of conditions.

Although it is broadly true that life has evolved from 'easier' to more 'difficult' environments, and that living on land is in this sense more hazardous than living in water, it would be a mistake to press the principle too far, particularly in the case of terrestrial arthropods, whose forte is specialization. Arthropods which live in deserts are indeed well adapted to difficult terrestrial conditions, and their adaptations are interesting to explore, but they are not necessarily more advanced in the sense of being independent of the environment over a wider range, for they cannot survive in moist, cold surroundings. Adaptability is a property of the group as a whole rather than of individual species.

REFERENCES

ALEXANDER, P., KITCHENER, J. A. and BRISCOE, H. V. A. (1944). Inert dust insecticides. *Ann. appl. Biol.* **31**, 143–59.

AUZOU, M. L. (1953). Recherches biologiques et physiologiques sur deux isopodes onisciens: *Porcellio scaber* Lat. et *Oniscus asellus* L. *Ann. Sci. nat.* (*Zool.*), **15**, 71–98.

BABCOCK, S. M. (1912). Metabolic water: its production and role in vital phenomena. *Res. Bull. Wis. agric. Exp. Sta.* no. 22, pp. 87–181.

BACHMAN, E. L. (1912). Der osmotische Druck bei einiger Wasserkafern. *Arch. ges. Physiol.* **149**, 93–114.

BACHMETJEW, P. (1899). Über die Temperatur der Insekten nach Beobachtungen in Bulgarien. *Z. wiss. Zool.* **66**, 520–604.

BARRER, R. M. (1939). Permeation, diffusion and solution of gases in organic polymers. *Trans. Faraday Soc.* **35**, 628–43.

BARRER, R. M. (1941). *Diffusion in and through solids.* Cambridge University Press.

BEADLE, L. C. (1939). Regulation of the haemolymph in the saline water mosquito larva *Aëdes detritus* Edw. *J. exp. Biol.* **16**, 346–62.

BEADLE, L. C. and SHAW, J. (1950). The retention of salt and the regulation of the non-protein nitrogen fraction in the blood of the aquatic larva, *Sialis lutaria. J. exp. Biol.* **27**, 96–109.

BEAMENT, J. W. L. (1945). The cuticular lipoids of insects. *J. exp. Biol.* **21**, 115–31.

BEAMENT, J. W. L. (1946a). The waterproofing process in eggs of *Rhodnius prolixus* Stahl. *Proc. roy. Soc.* B, **133**, 407–18.

BEAMENT, J. W. L. (1946b). The formation and structure of the chorion of the egg in an hemipteran, *Rhodnius prolixus. Quart. J. micr. Sci.* **87**, 393–439.

BEAMENT, J. W. L. (1948). The role of wax layers in the waterproofing of insect cuticle and egg-shell. *Disc. Faraday Soc.* no. 3, pp. 177–82.

BEAMENT, J. W. L. (1949). The penetration of insect egg-shells. II. The properties and permeability of the sub-chorial membranes during development of *Rhodnius prolixus* Stahl. *Bull. ent. Res.* **39**, 467–88.

BEAMENT, J. W. L. (1951). The structure and formation of the egg of the fruit-tree red spider mite *Metatetranychus ulmi* Koch. *Ann. appl. Biol.* **38**, 1–24.

BEAMENT, J. W. L. (1954). Water transport in insects. *Symp. Soc. exp. Biol.* **8**, 94–117.

BEAMENT, J. W. L. (1955). Wax secretion in the cockroach. *J. exp. Biol.* (in the press).

BEATTIE, M. V. F. (1928). Observations on the thermal death points of the blowfly at different relative humidities. *Bull. ent. Res.* **18**, 397–403.

94

BERGER, B. (1907). Über die Widerstandsfähigkeit der Tenebriolarven gegen Austrocung. *Arch. ges. Physiol.* **118**, 607–12.

BERGMANN, W. (1938). The composition of the ether extractives from exuviae of the silkworm, *Bombyx mori. Ann. ent. Soc. Amer.* **31**, 315–21.

BIRCH, L. C. and ANDREWARTHA, H. G. (1942). The influence of moisture on the eggs of *Austroicetes cruciata* Sauss. (Orthoptera) with reference to their ability to survive desiccation. *Aust. J. exp. Biol. med. Sci.* **20**, 1–8.

BLOWER, G. A. (1951). Comparative study of the chilopod and diplopod cuticle. *Quart. J. micr. Sci.* **92**, 141–61.

BODENHEIMER, F. S. (1935). *Animal life in Palestine.* Jerusalem.

BODENHEIMER, F. S. and SAMBURSKI, K. (1930). Ueber den Wärmeausgleich bei Insekten. *Zool. Anz.* **86**, 208–11.

BODINE, J. H. (1921). Factors influencing the water content and the rate of metabolism of certain Orthoptera. *J. exp. Zool.* **32**, 137–64.

BODINE, J. H. (1923). Hibernation in Orthoptera. I. Physiological changes during hibernation in certain Orthoptera. *J. exp. Zool.* **37**, 457–76.

BONÉ, G. J. (1943). Recherches sur les glandes coxales et la régulation du milieu interne chez l'*Ornithodorus moubata* Murray (Acarina, Ixodidae). *Ann. Soc. zool. Belg.* **74**, 16–31.

BONÉ, G. J. (1945). Le rapport sodium-potassium dans le liquide coelomique des insectes. *Ann. Soc. zool. Belg.* **75**, 123–32.

BONÉ, G. J. (1947). Regulation of the sodium-potassium ratio in insects. *Nature, Lond.*, **160**, 679–80.

BONÉ, G. J. and KOCH, H. J. (1942). Le rôle des tubes de Malpighi et du rectum dans la régulation ionique chez les insectes. *Ann. Soc. zool. Belg.* **73**, 73–87.

BREITENBRECHER, J. K. (1918). The relation of water to the behaviour of the potato beetle in the desert. *Publ. Carneg. Instn*, no. 263, pp. 341–84.

BROWNING, T. O. (1953). The influence of temperature and moisture on the uptake and loss of water in the eggs of *Gryllulus commodus* Walker (Orthoptera-Gryllidae). *J. exp. Biol.* **30**, 104–15.

BROWNING, T. O. (1954a). On the structure of the spiracle of the tick, *Ornithodorus moubata* Murray. *Parasitology*, **44**, 310–12.

BROWNING, T. O. (1954b). Water balance in the tick *Ornithodorus moubata* Murray, with particular reference to the influence of carbon dioxide on the uptake and loss of water. *J. exp. Biol.* **31**, 331–40.

BUDDENBROCK, W. v. and ROHR, G. v. (1922). Die Atmung von *Dixippus morosus. Z. allg. Physiol.* **20**, 111–60.

BUISSON, M. DE (1928). Recherches sur la ventilation trachéenne chez les chilopodes et sur la circulation sanguine chez les scutigères. *Arch. Zool. exp. gén.* **67**, 49–63.

BURSELL, E. (1955). The transpiration of terrestrial isopods. *J. exp. Biol.* **32**, 238–55.

BUXTON, P. A. (1924a). Heat, moisture and animal life in deserts. *Proc. roy. Soc. B*, **96**, 123–31.

BUXTON, P. A. (1924b). The temperature of the surface of deserts. *J. Ecol.* **12**, 127–34.

BUXTON, P. A. (1930). Evaporation from the mealworm and atmospheric humidity. *Proc. roy. Soc.* B, **106**, 560–77.

BUXTON, P. A. (1931*a*). The measurement and control of atmospheric humidity in relation to entomological problems. *Bull. ent. Res.* **22**, 431–47.

BUXTON, P. A. (1931*b*). The law governing the loss of water from an insect. *Proc. R. ent. Soc. Lond.* **6**, 27–31.

BUXTON, P. A. (1932*a*). Terrestrial insects and the humidity of the environment. *Biol. Rev.* **7**, 275–320.

BUXTON, P. A. (1932*b*). The relation of adult *Rhodnius prolixus* (Reduviidae, Rhyncota) to atmospheric humidity. *Parasitology*, **24**, 429–39.

BUXTON, P. A. and LEWIS, D. J. (1934). Climate and tsetse flies: laboratory studies upon *Glossina submorsitans* and *tachinoides*. *Phil. Trans.* B, **224**, 175–240.

CLAUS, A. (1937). Vergleichend-physiologische Untersuchungen sur Oekologie der Wasserwanzen mit besonderer Berucksichtigung der Brachwasser-wanze. *Zool. Jb.* (*Physiol.*), **58**, 365–432.

CLOUDSLEY-THOMPSON, J. L. (1950). The water relations and cuticle of *Paradesmus gracilis* (Diplopoda, Strongylidae). *Quart. J. micr. Sci.* **91**, 453–64.

COLOSI, I. DE S. (1933). L'assunzione dell'acqua per via cutanea. *Pubbl. Staz. zool. Napoli*, **13**, 12–38.

CRISP, D. J. and THORPE, W. H. (1948). The water-protecting properties of insect hairs. *Disc. Faraday Soc.* no. 3, pp. 210–20.

DAVIES, L. (1948). Laboratory studies on the egg of the blow-fly *Lucilia sericata*. *J. exp. Biol.* **25**, 71–85.

DAVIES, L. (1950). The hatching mechanism of muscid eggs. *J. exp. Biol.* **27**, 437–45.

DAVIES, M. E. and EDNEY, E. B. (1952). The evaporation of water from spiders. *J. exp. Biol.* **29**, 571–82.

DAVIES, W. M. (1928). The effects of variation in relative humidity on certain species of Collembola. *J. exp. Biol.* **6**, 79–86.

DECKER, G. C. and ANDRÉ, F. (1936). Studies on temperature and moisture as factors influencing winter mortality in adult chinch bugs. *Iowa St. Col. J. Sci.* **10**, 403–20.

DELAUNAY, H. (1931). L'excrétion azotée des invertébrés. *Biol. Rev.* **31**, 265–301.

DENNELL, R. (1946). A study of an insect cuticle: the larval cuticle of *Sarcophaga falculata* Pand. (Diptera). *Proc. roy. Soc.* B, **133**, 348–73.

DENNELL, R. and MALEK, S. R. A. (1955). The cuticle of the cockroach *Periplaneta americana*. I–IV. *Proc. roy. Soc.* B, **143**, 126–36, 239–56, 414–34.

DIGBY, P. S. B. (1955). Factors affecting the temperature excess of insects in sunshine. *J. exp. Biol.* **32**, 279–98.

DITMAN, L. P., VOGT, G. B. and SMITH, D. R. (1942). The relation of unfreezable water to cold hardiness of insects. *J. econ. Ent.* **35**, 265–72.

DOTY, P. M., AIKEN, W. H. and HERMANN, M. (1944). Water vapour permeability of organic films. *Industr. Engng Chem.* (*Anal.*), **16**, 686–90.

Dresel, I. B. and Moyle, V. (1950). Nitrogenous excretion of amphipods and isopods. *J. exp. Biol.* **27**, 210–25.

Duval, M., Portier, P. and Courtois, A. (1928). Sur la présence de grandes quantités d'acides amines dans le sang des Insectes. *C.R. Acad. Sci., Paris*, **186**, 652–3.

Eder, R. (1940). Die kutikuläre Transpiration der Insekten und ihre Abhängigkeit vom Aufbau des Integuments. *Zool. Jb.* **60**, 203–40.

Edney, E. B. (1945). Laboratory studies on the bionomics of the rat fleas, *Xenopsylla brasiliensis* Baker and *X. cheopsis* Roths. I. Certain effects of light, temperature and humidity on the rate of development and on adult longevity. *Bull. ent. Res.* **35**, 399–416.

Edney, E. B. (1947). Laboratory studies on the bionomics of the rat fleas, *Xenopsylla brasiliensis* Baker and *X. cheopsis* Roths. II. Water relations during the cocoon period. *Bull. ent. Res.* **38**, 263–80.

Edney, E. B. (1949). Evaporation of water from woodlice. *Nature, Lond.*, **164**, 321.

Edney, E. B. (1951a). The evaporation of water from woodlice and the millipede *Glomeris*. *J. exp. Biol.* **28**, 91–115.

Edney, E. B. (1951b). The body temperature of woodlice. *J. exp. Biol.* **28**, 271–80.

Edney, E. B. (1952). The body temperature of arthropods. *Nature, Lond.*, **170**, 586–7.

Edney, E. B. (1953). The temperature of woodlice in the sun. *J. exp. Biol.* **30**, 331–49.

Edney, E. B. (1954). Woodlice and the land habitat. *Bio. Rev.* **29**, 185–219.

Edney, E. B. and Spencer, J. O. (1955). Cutaneous respiration in woodlice. *J. exp. Biol.* **32**, 256–69.

Evans, A. C. (1944). Observations on the biology and physiology of wireworms of the genus *Agriotes* Esch. *Ann. appl. Biol.* **31**, 235–50.

Florkin, M. and Duchateau, G. (1943). Les formes du système enzymatique de l'uricolyse et l'évolution du catabolisme purique chez les animaux. *Arch. int. Physiol.* **53**, 267–307.

Folco, G. B. (1932). Contributo alla conoscenza dei tube di Malpighi di *Pachyiulus communis* (Savi). *P.V. Soc. tosc. Sci. nat.* **41**, 93–8.

Fraenkel, G. (1930). Der Atmungsmechanismus der Scorpions. Ein Beitrag zur Physiologie der Tracheenlunge. *Z. vergl. Physiol.* **11**, 656–61.

Fraenkel, G. and Blewett, M. (1944). The utilisation of metabolic water in insects. *Bull. ent. Res.* **35**, 127–39.

Fraenkel, G. and Herford, G. V. B. (1938). The respiration of insects through the skin. *J. exp. Biol.* **15**, 266–80.

Gibbins, E. G. (1932). A note on the relative size of the anal gills of mosquito larvae breeding in salt and fresh water. *Ann. trop. Med. Parasit.* **26**, 551–4.

Gorvett, H. (1950). 'Weber's glands' and respiration in woodlice. *Nature, Lond.*, **166**, 115.

Govaerts, J. and Leclercq, J. (1946). Water exchange between insects and air moisture. *Nature, Lond.*, **157**, 483.

H

GUNN, D. L. (1933). The temperature and humidity relations of the cockroach, *Blatta orientalis*. I. Desiccation. *J. exp. Biol.* **10**, 274–85.

GUNN, D. L. (1942). Body temperature in poikilothermal animals. *Biol. Rev.* **17**, 293–314.

GUNN, D. L. and COSWAY, C. A. (1942). The temperature and humidity relations of the cockroach. VI. Oxygen consumption. *J. exp. Biol.* **19**, 124–32.

GUBB, D. L. and NOTLEY, F. B. (1936). The temperature and humidity relations of the cockroach. IV. Thermal death point. *J. exp. Biol.* **13**, 28–34.

HARTLEY, G. S. (1948). Communication. *Disc. Faraday Soc.* no. 3, p. 223.

HAZELHOFF, E. H. (1926). *Regeling der Ademholing bij Insecten en Spinnen.* Utrecht.

HENSON, H. (1937). The structure and post embryonic development of *Vanessa urticae* (Lep.). II. The larval Malpighian tubules. *Proc. zool. Soc. Lond.* B, **107**, 161–74.

HERS, M. J. (1942). Anaerobiose et régulation minérale chez les larves de *Chironomus*. *Ann. Soc. zool. Belg.* **73**, 173–9.

HODSON, A. C. (1937). Some aspects of the role of water in insect hibernation. *Ecol. Monogr.* **7**, 271–315.

HOLDGATE, M. W. (1955*a*). The wetting and water-proofing properties of some insect cuticles. Thesis, Cambridge University.

HOLDGATE, M. W. (1955*b*). The wetting of insect cuticles by water. *J. exp. Biol.* **32**, 591–617.

HOLDGATE, M. W. (1956). Transpiration through the cuticles of some aquatic insects. *J. exp. Biol.* **33**, 107–18.

HOLDGATE, M. W., MENTER, J. W. and SEAL, M. (1955). A study of an insect surface by reflection electron microscopy and diffraction. *Proc. 2nd Int. Conference on Electron Miscroscopy*, R. micr. Soc. London.

HOLDGATE, M. W. and SEAL, M. (1956). The epicuticular wax layers in the pupa of *Tenebrio molitor* (L.). *J. exp. Biol.* **33**, 82–106.

HURST, H. (1940). Permeability of insect cuticle. *Nature, Lond.*, **145**, 462–3.

HURST, H. (1941). Insect cuticle as an asymmetrical membrane. *Nature, Lond.*, **147**, 388–9.

HURST, H. (1948). Asymmetrical behaviour of insect cuticle in relation to water permeability. *Disc. Faraday Soc.* no. 3, pp. 193–210.

JOHNSON, C. G. (1937). The absorption of water and associated volume changes occurring in the eggs of *Notostira erratica* L. (Hemiptera, Capsidae) during embryonic development under experimental conditions. *J. exp. Biol.* **14**, 413–21.

JOHNSON, C. G. (1942). Insect survival in relation to the rate of water loss. *Biol. Rev.* **17**, 151–77.

JONES, G. D. G. (1955). The cuticular waterproofing mechanism of the worker honey-bee. *J. exp. Biol.* **32**, 95–109.

JONES, S. E. (1941). Influence of temperature and humidity on the life history of the spider *Agelena naevia* Walckenaer. *Ann. ent. Soc. Amer.* **34**, 557–71.

JORDAAN, H. (1927). Die Regulierung der Atmung bei Insekten und Spinnen. *Z. vergl. Physiol.* **5**, 179–90.

KALMUS, H. (1936). Die Verwendung der Tracheenblasen der Corethra-larve als Mikrohygrometer. *Z. wiss. Mikr.* **53**, 215–19.

KALMUS, H. (1941). Physiology and ecology of cuticle colour in insects. *Nature, Lond.*, **148**, 428–31.

KASTNER, A. (1929). Bau und Funktion der Fächertracheen einiger Spinnen. *Z. Morph. Ökol. Tiere*, **13**, 463–558.

KASTNER, A. (1935). Die Funktion der sogenannten sympathischen Ganglien und die Excretion bei den Phalangiiden. *Zool. Anz.* **109**, 273–88.

KOCH, H. J. (1938). The absorption of chloride ions by the anal papillae of Diptera larvae. *J. exp. Biol.* **15**, 152–60.

KOIDSUMI, K. (1935). Experimentelle Studien uber die Transpiration und den Wärmehaushalt bei Insekten. *Mem. Fac. Sci. Agric. Taihoku*, **12**, 1–380.

KRAMER, S. and WIGGLESWORTH, V. B. (1950). The outer layers of the cuticle in the cockroach *Periplaneta americana* and the function of the oenocytes. *Quart. J. micr. Sci.* **91**, 63–72.

KROGH, A. (1919). The diffusion of gases through animal tissues. *J. Physiol.* **52**, 391–408.

KROGH, A. (1939). *Osmotic regulation in aquatic animals*. Cambridge University Press.

LAFON, M. (1943). Recherches biochimiques et physiologiques sur le squelette tégumentaire des arthropodes. *Ann. Sci. nat. (Zool.)* (11), **5**, 113–46.

LAFON, M. and TEISSIER, G. (1939). Inanition et metamorphose chez *Tenebrio molitor*. *C.R. Soc. Biol., Paris*, **131**, 417.

LAUGHLIN, R. (1953). Absorption of water by the egg of the garden chafer *Phyllopertha horticola*. *Nature, Lond.*, **171**, 577.

LECLERCQ, J. (1946). Des insectes qui boivent de l'eau. *Bull. Ann. Soc. ent. Belge*, **82**, 71–5.

LECLERCQ, J. (1948). Influence des conditions hygrométriques sur les œufs de *Melasoma populi* L. (Col. Chrysomelidae). *Bull. Ann. Soc. ent. Belge*, **84**, 26–7.

LEES, A. D. (1946a). The water balance in *Ixodes ricinus* L. and certain other species of ticks. *Parasitology*, **37**, 1–20.

LEES, A. D. (1946b). Chloride regulation and the function of the coxal glands in ticks. *Parasitology*, **37**, 172–84.

LEES, A. D. (1947). Transpiration and the structure of the epicuticle in ticks. *J. exp. Biol.* **23**, 379–410.

LEES, A. D. and BEAMENT, J. W. L. (1948). An egg-waxing organ in ticks. *Quart. J. micr. Sci.* **89**, 291–332.

LESTER, H.M.O. and LLOYD, L. (1928). Notes on the process of digestion in tsetse-flies. *Bull. ent. Res.* **19**, 39–60.

LICENT, P. E. (1912). Recherches d'anatomie et de physiologie comparée sur le tube digestif des homoptères supérieurs. *La Cellule*, **28**, 7–161.

99

Lison, L. (1938). Contribution à l'étude morphologique et histophysiologique du système malpighien de *Melolontha melolontha* Linn. (Coleoptera). *Ann. Soc. zool. Belg.* **69**, 195–233.

Ludwig, D. (1937). The effect of different relative humidities on respiratory metabolism and survival of the grasshopper *Chortophaga viridifasciata* de Geer. *Physiol. Zoöl.* **10**, 342–51.

Ludwig, D. (1945). The effects of atmospheric humidity on animal life. *Physiol. Zoöl.* **18**, 103–35.

Ludwig, D. and Wugmeister, M. (1953). Effects of starvation on the blood of the Japanese beetle (*Popillia japonica* Newman) larvae. *Physiol. Zoöl.* **26**, 254–9.

Maloeuf, N. S. R. (1938). Physiology of excretion among the Arthropoda. *Physiol. Rev.* **18**, 28–58.

Manton, S. M. and Heatley, N. G. (1937). Studies on the Onycophora— II. The feeding, digestion, excretion, and food storage of *Peripatopsis*. *Phil. Trans.* B, **227**, 411–64.

Manton, S. M. and Ramsay, J. A. (1937). Studies on the Onycophora— III. The control of water loss in *Peripatopsis*. *J. exp. Biol.* **14**, 470–2.

Matthée, J. J. (1950). The structure and physiology of the egg of *Locusta pardalina* (Walk.). *Abstr. Diss. Univ. Camb.*, 1948–9, pp. 29–30.

Matthée, J. J. (1951). The structure and physiology of the egg of *Locusta pardalina* (Walk.). *Sci. Bull. Dep. Agric. S. Afr.* no. 316, pp. 1–83.

Mead-Briggs, A. R. (1954). Studies on water relations of arthropods. Thesis, Birmingham University.

Meinertz, T. (1944). Beiträge zur Ökologie der Landisopoden mit besonderer Berücksichtigung ihrer Atmungsorgane. *Zool. Jb.* (*Syst.*), **76**, 501–18.

Mellanby, K. (1932a). The effect of atmospheric humidity on the metabolism of the fasting meal-worm (*Tenebrio molitor* L., Coleoptera). *Proc. roy. Soc.* B, **111**, 376–90.

Mellanby, K. (1932b). Effects of temperature and humidity on the metabolism of the fasting bed-bug (*Cimex lectularius*), Hemiptera. *Parasitology*, **24**, 419–28.

Mellanby, K. (1932c). The influence of atmospheric humidity on the thermal death points of a number of insects. *J. exp. Biol.* **9**, 222–31.

Mellanby, K. (1934a). Effects of temperature and humidity on the clothes moth larva, *Tineola Biselliella* Hum. (Lepidoptera). *Ann. appl. Biol.* **21**, 476–82.

Mellanby, K. (1934b). The site of water loss from insects. *Proc. roy. Soc.* B, **116**, 139–49.

Mellanby, K. (1935a). The evaporation of water from insects. *Biol. Rev.* **10**, 317–33.

Mellanby, K. (1935b). The structure and function of the spiracles of the tick, *Ornithodorus moubata* Murray. *Parasitology*, **27**, 288–90.

Mellanby, K. (1939a). The functions of insect blood. *Biol. Rev.* **14**, 243–60.

Miley, H. H. (1930). Internal anatomy of *Euryurus erythropygus* (Brandt) (Diplopoda). *Ohio J. Sci.* **30**, 229–49.

MILLER, A. (1940). Embryonic membranes, yolk cells and morphogenesis of the stonefly *Pteronarcys proteus* Newman. *Ann. ent. Soc. Amer.* **33**, 437–77.

MILLOT, J. (1926). Contribution à l'histophysiologie des aranéides. *Bull. biol.* Suppl. 8, 1–238.

MILLOT, J. (1942). Sur l'anatomie et l'histophysiologie de *Koenenia mirabilis* (Arachnida, Palpigradi). *Rev. franc. Ent.* **9**, 33–51.

MILLOT, J. and FONTAINE, M. (1937). La teneur en eau des aranéides. *Bull. Soc. zool. Fr.* **62**, 113–19.

MILLOT, J. and PAULIAN, R. (1943). Valeur fonctionelle de poumons des scorpions. *Bull. Soc. zool. Fr.* **68**, 97–8.

MÖDLINGER, G. (1931). Morphologie der Respirationsorgane der Landiso-poden. *Studia Zool.* **2**, 25–79.

MÖDLINGER, G. (1934). Beiträge zur Histologie der Isopoden. *Allatt. Közlem.* **31**, 42–7.

MORRISON, P. R. (1946). Physiological observations on water loss and oxygen consumption in *Peripatus*. *Biol. Bull.*, Woods Hole, **91**, 181–8.

MUNSON, S. C. and YEAGER, J. F. (1949). Blood volume and chloride normality in roaches (*Periplaneta americana* (L.)) injected with sodium chloride solutions. *Ann. ent. Soc. Amer.* **42**, 165–73.

NECHELES, H. (1924). Über Wärmeregulation bei wechselwarmen Tiere. *Arch. ges. Physiol.* **204**, 72–93.

NEEDHAM, J. (1929). Protein metabolism and organic evolution. *Science Progress*, **23**, 633–46.

NEEDHAM, J. (1935). Problems of nitrogen catabolism in invertebrates. II. Correlation between uricotelic metabolism and habitat in the phylum Mollusca. *Biochem. J.* **29**, 238–51.

NUTMAN, S. R. (1941). Function of the ventral tube in *Onychiurus armatus* (Collembola). *Nature, Lond.*, **148**, 168–9.

PAL, R. (1950). The wetting of insect cuticle. *Bull. ent. Res.* **41**, 121–39.

PARRY, D. A. (1951). Factors determining the temperature of terrestrial arthropods in sunlight. *J. exp. Biol.* **28**, 445–62.

PARRY, D. A. (1954). On the drinking of soil capillary water by spiders. *J. exp. Biol.* **31**, 218–27.

PARRY, G. (1953). Osmotic and ionic regulation in the isopod crustacean *Ligia oceanica*. *J. exp. Biol.* **30**, 567–74.

PATAY, R. (1938). Sur la structure et l'histophysiologie des tubes de Malpighi chez le doryphore. *C.R. Soc. Biol.*, Paris, **129**, 1098–9.

PATTON, R. L. and CRAIG, R. (1939). The rate of excretion of certain sub-stances by the larvae of the mealworm, *Tenebrio molitor* L. *J. exp. Zool.* **81**, 437–51.

PAYNE, N. M. (1927a). Measures of insect cold hardiness. *Biol. Bull.*, Woods Hole, **52**, 449–57.

PAYNE, N. M. (1927b). Freezing and survival of insects at low tempera-tures. *J. Morph.* **43**, 521–46.

PESCHEN, K. E. (1939). Untersuchungen über das Vorkommen und den Stofwechsel des Guanins im Tierreich. *Zool. Jb.* (*Allg. Zool.*), **59**, 429–62.

PICKEN, L. E. R. (1936). The mechanism of urine formation in invertebrates. I. The excretion mechanism in certain arthropods. *J. exp. Biol.* **13**, 309–28.

POLL, M. (1938). Contribution à l'étude de l'appareil urinaire des chenilles de lépidoptères. *Ann. Soc. zool. Belg.* **69**, 9–52.

PORTIER, P. and DUVAL, M. (1927). Concentration moléculaire et teneur en chlore du sang de quelques insectes. *C.R. Soc. Biol., Paris*, **95**, 1605–6.

PROSSER, C. L. (ed.) (1952). *Comparative animal physiology.* London: Saunders.

PRYOR, M. G. M. (1940). On the hardening of the cuticle of insects. *Proc. roy. Soc.* B, **128**, 393–407.

RAMSAY, J. A. (1935*a*). Methods of measuring the evaporation of water from animals. *J. exp. Biol.* **12**, 355–72.

RAMSAY, J. A. (1935*b*). The evaporation of water from the cockroach. *J. exp. Biol.* **12**, 373–83.

RAMSAY, J. A. (1949). A new method of freezing point determination for small quantities. *J. exp. Biol.* **26**, 57–74.

RAMSAY, J. A. (1950). Osmotic regulation in mosquito larvae. *J. exp. Biol.* **27**, 145–57.

RAMSAY, J. A. (1951). Osmotic regulation in mosquito larvae; the role of the Malpighian tubules. *J. exp. Biol.* **28**, 62–73.

RAMSAY, J. A. (1952). The excretion of sodium and potassium by the Malpighian tubule of *Rhodnius*. *J. exp. Biol.* **29**, 110–26.

RAMSAY, J. A. (1953*a*). Exchanges of sodium and potassium in mosquito larvae. *J. exp. Biol.* **30**, 79–89.

RAMSAY, J. A. (1953*b*). Active transport of potassium by the Malpighian tubules of insects. *J. exp. Biol.* **30**, 358–69.

RAMSAY, J. A. (1954). Active transport of water by the Malpighian tubules of the stick insect *Dixippus morosus* (Orthoptera, Phasmidae). *J. exp. Biol.* **31**, 104–13.

RAMSAY, J. A., BROWN, R. H. J. and FALLOON, S. W. H. W. (1953). Simultaneous determination of sodium and potassium in small volumes of fluid by flame photometry. *J. exp. Biol.* **30**, 1–17.

REMY, P. (1925). *Contributions à l'étude de l'appareil respiratoire et de la respiration chez quelques invertébrés.* Nancy: Wagner.

RICHARDS, A. G. (1951). *The integument of arthropods.* Minneapolis: University of Minnesota Press.

RICHARDS, A. G. and ANDERSON, T. F. (1942). Electron microscope studies of insect cuticle with a discussion of the application of electron optics to this problem. *J. Morph.* **71**, 135–83.

RICHARDS, A. G., CLAUSEN, M. B. and SMITH, M. N. (1953). Studies on arthropod cuticle. X. The asymmetrical penetration of water. *J. cell. comp. Physiol.* **42**, 395–413.

ROBINSON, W. (1928*a*). Relation of hydrophilic colloids to cold hardiness in insects. *Colloid Symposium Monogr.* **5**, 199–218.

ROBINSON, W. (1928*b*). Response and adaptation of insects to external stimuli. *Ann. ent. Soc. Amer.* **21**, 407–17.

SACCHAROV, N. L. (1930). Studies in cold resistance of insects. *Ecology*, **11**, 505–17.

SALT, R. W. (1936). Studies on the freezing process in insects. *Tech. Bull. Minn. agric. Exp. Sta.* **116**, 1–41.

SALT, R. W. (1949). Water uptake in eggs of *Melanoplus bivattatus* (Say). *Canad. J. Res.* D, **27**, 236–42.

SALT, R. W. (1952). Some aspects of moisture absorption and loss in eggs of *Melanoplus bivattatus* (Say). *Canad. J. Zool.* **30**, 55–82.

SALT, R. W. (1953). The influence of food on cold hardiness of insects. *Canad. Ent.* **85**, 261–9.

SCHALLER, F. (1949). Osmoregulation und Wasserhaushalt der Larve von *Corethra plumicornis*, mit besonderer Berücksichtigung der Vorgänge am Darmkanal. *Z. vergl. Physiol.* **31**, 684–95.

SCHNEIDER, F. (1948). Beitrag zur Kenntnis der Generationsverhältnisse und Diapause räuberischer Schwebfliegen (Syrphidae, Dipt.). *Mitt. schweiz. ent. Ges.* **21**, 249–85.

SCHULTZ, F. N. (1930). Zur Biologie des Mehlwurms (*Tenebrio molitor*). I. Der Wasserhaushalt. *Biochem. Z.* **127**, 112–19.

SHAW, J. (1955a). The permeability and structure of the cuticle of the aquatic larva of *Sialis lutaria*. *J. exp. Biol.* **32**, 330–52.

SHAW, J. (1955b). Ionic regulation and water balance in the aquatic larva of *Sialis lutaria*. *J. exp. Biol.* **32**, 353–82.

SIEGLER, E. H. (1946). Susceptibility of hibernating codling moth larvae to low temperatures, and the bound water content. *J. agric. Res.* **72**, 329–40.

SLIFER, E. H. (1938). The formation and structure of a special water absorbing area in the membranes covering the grasshopper egg. *Quart. J. micr. Sci.* **80**, 437–57.

SLIFER, E. H. (1946). The effects of xylol and other solvents on diapause in the grasshopper egg; together with a possible explanation for the action of these agents. *J. exp. Zool.* **102**, 333–56.

SLIFER, E. H. (1954). The permeability of the sensory pegs on the antennae of the grasshopper. *Biol. Bull., Woods Hole*, **106**, 128–18.

SLIFER, E. H. (1955). The distribution of permeable sensory pegs on the body of the grasshopper. *Ent. News*, **66**, 1–5.

SMART, J. (1935). The effect of temperature and humidity on the cheese skipper, *Piophila casei* (L.). *J. exp. Biol.* **12**, 384–8.

SPEICHER, B. R. (1931). The effects of desiccation upon the growth and development of the Mediterranean flour-moth. *Proc. Pa Acad. Sci.* **5**, 79.

SPENCER, J. O. and EDNEY, E. B. (1954). The absorption of water by woodlice. *J. exp. Biol.* **31**, 491–6.

THEODOR, O. (1936). On the relation of *Phlebotomus patatasi* to the temperature and humidity of the environment. *Bull. ent. Res.* **27**, 653–71.

THOMPSON, V. and BODINE, J. H. (1936). Oxygen consumption and rates of dehydration of grasshopper eggs. *Physiol. Zoöl.* **9**, 455–70.

THORPE, W. H. and CRISP, D. J. (1946). Studies on plastron respiration. I. The biology of *Aphelocheirus* and the mechanism of plastron retention. *J. exp. Biol.* **24**, 227–69.

TIEGS, O. W. (1942a). The 'dorsal organ' of collembolan embryos. *Quart. J. micr. Sci.* **83**, 153–69.

TIEGS, O. W. (1942b). The 'dorsal organ' of the embryo of *Campodea*. *Quart. J. micr. Sci.* **84**, 35–47.

TIMON-DAVID, J. (1945). Fragments de biochimie entomologiques. III. Excrétion et sécrétions. *Ann. Fac. Sci. Marseille*, **16**, 179–235.

TOBIAS, J. M. (1948). Potassium, sodium and water interchange in irritable tissues and haemolymph of an omnivorous insect, *Periplaneta americana*. *J. cell. comp. Physiol.* **31**, 125–42.

TREHERNE, J. E. (1954a). Osmotic regulation in the larvae of *Helodes* (Coleoptera—Helodidae). *Trans. R. ent. Soc. Lond.* **105**, 117–30.

TREHERNE, J. E. (1954b). The exchange of labelled sodium in the larva of *Aëdes aegypti* L. *J. exp. Biol.* **31**, 386–401.

UVAROV, B. P. (1931). Insects and climate. *Trans. R. ent. Soc. Lond.* **79**, 1–247.

UVAROV, B. P. (1948). Recent advances in acridology: anatomy and physiology of Acridiidae. *Trans. R. ent. Soc. Lond.* **99**, 1–75.

VERHOEFF, K. W. (1920). Über die Atmung der Landasseln, zugleich ein Beitrag zur Kenntnis der Entstehung der Landtiere. *Z. wiss. Zool.* **118**, 365–447.

VERHOEFF, K. W. (1941). Zur Kenntnis der Chilopodenstigmen. *Z. Morph. Ökol. Tiere*, **38**, 96–117.

VINOGRADSKAJA, O. N. (1936). Osmotische Druck der Haemolymph bei *Anopheles maculipennis messeae* Fell. *Z. Parasitenk.* **8**, 697–713.

WANG, T. H. and WU, H. W. (1948). On the structure of the Malpighian tubes of the centipede and their excretion of uric acid. *Sinensia Shanghai*, **18**, 1–11.

WATERHOUSE, F. L. (1951). Body temperature of small insect larvae. *Nature, Lond.*, **168**, 340.

WAY, M. J. (1950). The structure and development of the larval cuticle of *Diataraxia oleracea* (Lepidoptera). *Quart. J. micr. Sci.* **91**, 145–82.

WEBB-FOWLER, A. (1955). A preliminary study of moulting in isopods. Thesis, Birmingham University.

WEBER, H. (1930). *Biologie der Hemipteren*. Berlin.

WEBER, H. (1931). Lebensweise und Umweltbezeihungen von *Trialeurodes vaporariorum* (Aleurodina). *Z. Morph. Ökol. Tiere*, **23**, 575–753.

WIDMANN, E. (1935). Osmoregulation bei einheimischen Wasser und Feuchtluft-Crustaceen. *Z. wiss. Zool.* **147**, 132–69.

WIGGLESWORTH, V. B. (1931). The physiology of excretion in a blood-sucking insect, *Rhodnius prolixus* (Hemiptera, Reduviidae). *J. exp. Biol.* **8**, 411–51.

WIGGLESWORTH, V. B. (1932). On the function of the so-called 'rectal glands' of insects. *Quart. J. micr. Sci.* **75**, 131–50.

WIGGLESWORTH, V. B. (1933a). The physiology of the cuticle and of ecdysis in *Rhodnius prolixus*, with special reference to the functions of the oenocytes and of the dermal glands. *Quart. J. micr. Sci.* **76**, 270–318.

WIGGLESWORTH, V. B. (1933b). The effect of salts on the anal gills of the mosquito larva. *J. exp. Biol.* **10**, 1–15.

WIGGLESWORTH, V. B. (1933c). The function of the anal gills of the mosquito larva. *J. exp. Biol.* **10**, 16–26.

WIGGLESWORTH, V. B. (1938). The regulation of osmotic pressure and chloride concentration in the haemolymph of mosquito larvae. *J. exp. Biol.* **15**, 235–47.

WIGGLESWORTH, V. B. (1944). Action of inert dusts on insects. *Nature, Lond.*, **153**, 493–4.

WIGGLESWORTH, V. B. (1945). Transpiration through the cuticle of insects. *J. exp. Biol.* **21**, 97–114.

WIGGLESWORTH, V. B. (1947). The epicuticle of an insect, *Rhodnius prolixus* (Hemiptera). *Proc. roy. Soc.* B, **134**, 163–81.

WIGGLESWORTH, V. B. (1948a). The insect cuticle. *Biol. Rev.* **23**, 408–51.

WIGGLESWORTH, V. B. (1948b). The structure and deposition of the cuticle in the adult mealworm, *Tenebrio molitor* L. *Quart. J. micr. Sci.* **89**, 197–217.

WIGGLESWORTH, V. B. (1950). *The principles of insect physiology*, 4th ed. London: Methuen.

WIGGLESWORTH, V. B. (1953). Surface forces in the tracheal system of insects. *Quart. J. micr. Sci.* **94**, 507–22.

WIGGLESWORTH, V. B. and GILLETT, J. D. (1936). The loss of water during ecdysis in *Rhodnius prolixus* Stahl. (Hemiptera). *Proc. R. ent. Soc. Lond.* A, **11**, 104–7.

WILLIAMS, C. B. (1923). A short bioclimatic study in the Egyptian desert. *Bull. Minist. Agric. Egypt*, no. 29, pp. 1–20.

WILLIAMS, C. B. (1924a). Bioclimatic observations in the Egyptian desert in March 1923. *Bull. Minist. Agric. Egypt*, no. 37, pp. 1–18.

WILLIAMS, C. B. (1924b). A third bioclimatic study in the Egyptian desert. *Bull. Minist. Agric. Egypt*, no. 50, pp. 1–32.

YEAGER, J. F. and MUNSON, S. C. (1950). Blood volume of the roach *Periplaneta americana* determined by several methods. *Arthropoda*, **1**, 255–65.

ZOOND, A. (1931). Studies in the localisation of respiratory exchange in invertebrates. III. The book lungs of the scorpion. *J. exp. Biol.* **8**, 263–6.

ZOOND, A. (1934). The localisation of respiratory exchange in the scorpion. *Trans. roy. Soc. S. Afr.* **22**, xviii.

INDEX

Abrasion, effect of, on transpiration, 13–14, 27
abrasives, adsorption of, 14
Acridium, 52
Aëdes, 41–2, 47
A. aegypti, 41, 42, 43–6
Aeschna, 42
Agelena, 68
Agriotes, 17, 52–3, 86
air, drying power of, 2
air movement, effect on transpiration, 5, 79
anal drinking, 51–2
Anopheles, 39, 43–4, 49, 50, 74
Aphelocheirus, 29, 30
Armadillidium, 6, 37, 76–7, 80
Asellus, 37
Austroicetes, 70–1

Bibio, 17
bioclimates, 77–81
Blatta, 8, 14, 18, 76, 80
Blattella, 17, 22
Bombyx, 8
Buthus, 7

Calliphora, 60
Callosamia, 84
carbon dioxide, effect on transpiration, 7–8
cement, 13, 27–8
Chaerocampa, 9
Chironomus, 42, 50
chloride uptake, 41–1
chorion, 65
Chortophaga, 56–7, 83–4
Cimex, 38, 51, 54
cold-hardiness, 81 ff.
conduction, 79
convexion, 79
Corethra, 42
coxal fluid, 48

coxal glands, 37, 38
'critical' temperature, 16–17, 60, 89
cryptonephridia, 36
Culex, 39, 42–3
cutaneous respiration, 7 ff.
cuticle
    absorption of water through, 52–3
    asymmetrical permeability, 60 ff.
    general organization, 9, 13
    permeability to salts, 41 ff.
    permeability to water, 10, 19, 52–3
    wetting by water, 28
Cydia, 84

Dermestes, 54
Diacrisia, 83
Diataraxia, 18
diffusion, effect on transpiration, 18
Dixippus, 9, 47
drinking, 51 ff.
drying power of air, 2, 21
Dytiscus, 41

Eggs, water relations of, 65 ff.
Ephestia, 51, 54, 84, 86
epicuticle
    of arthropods, 10, 13
    of insects, 11
epicuticular waxes, 11, 18, *passim*
epidermal cells, secretory activity of, 58–9
Epistrophe, 52
evaporation, physics of, 3–4
excretion, 32, 87
    in arthropods other than insects, 36 ff.
    in insects, 33 ff.
excretory products, 32, 36, 87

Filter chamber, 36

Gain of water, *see* water absorption

affecting permeability of cuticle, 14 ff.
'critical', 16–17, 60, 89
lethal, 73 ff.
resistance to low, 81 ff.
transition, 16–17, 60, 89
temperature/transpiration curves, 15 ff.,
  21, 66 ff.
in constant saturation deficit, 19 ff.
interpretation, of 24–6, 66, 90
temperature in the field
effect of insolation on, 79–80
effect of transpiration on, 79
Tenebrio, 9, 11, 17–21, 26–7, 29, 30, 35,
  50, 51, 53, 55–6, 64, 74–5, 90–1
Tineola, 10, 54, 56
Tipula, 17
transition temperature, 16–17, 60, 89
transpiration, 6
and air movement, 5, 79
in aquatic insects, 17, 23
in arthropods generally, 7, 14–15,
  17–18, 21–3, 57
at constant saturation deficit, 19 ff.
and temperature, 18–20, 24, 66, 73,
  89
during moult, 12
effect of carbon dioxide on, 7–8
effect of death on, 58–9
effect of diffusion on, 18
exponential rise with temperature,
  25 ff., 90
from respiratory surfaces, 6–9, 87
Trialeurodes, 68
Tribolium, 51, 54

Undercooling point, 82 ff.

ureotely, 32
uricotely, 32, 38
urine, composition of, 39, 40, 46

Vanessa, 36

Water absorption
in ampullae, 33 ff., 50
through cuticle, 52–3, 55
by drinking, 51
by eggs, 67 ff.
with food, 51
from hypertonic environment, 49,
  91
from moist surfaces, 52, 91
in rectum, 33 ff.
water content, regulation of, 54–5
water loss
from respiratory surfaces, 6 ff.
from spiracles of insects, 8
water vapour absorption, 55–8
mechanism of, 65, 91
wax
in epicuticle, 13
in eggs, 66 ff.
distribution, of 28, 89
orientation of molecules of, 18
question of pores in, 25, 62–4, 89–
  90
repair, 27
solvents, 17, 26
waterproofing properties of, **11–12,**
  18, 80
Weber's glands, 6

Xenopsylla, 8–9, 58–9, 74, 77